Revise A2 Biology for AQA S[illegible]

Graham Read and Ray Skwi[illegible]

Heinemann Educational Publishers
Halley Court, Jordan Hill, Oxford, OX2 8EJ
Part of Harcourt Education

Heinemann is the registered trademark of Harcourt Education Limited

First published 2003

07 06 05 04 03
10 9 8 7 6 5 4 3 2 1

British Library Cataloguing in Publication Data is available from the British Library on request

ISBN 0 435 58307 7

Development editor Paddy Gannnon

Edited by Alexandra Clayton

Designed and typeset by Saxon Graphics Ltd, Derby

Index compiled by Diana Boatman

Printed and bound in Great Britain by Thomson Litho Ltd, Glasgow

Tel: 01865 888058 www.heinemann.co.uk

Contents

Introduction – How to use this revision guide iv
AQA A2 Biology – Assessment v

Module 5: Inheritance, Evolution and Ecosystems B HB

Introduction **2**
Transmission of genetic information
Meiosis (and Inheritance) 4
Principles of Mendelian Inheritance 6
Dihybrid crosses and chi-squared 8
Investigating variation
Investigating variation 10
Causes of variation 12
Hardy-Weinberg Principle 14
Selection and speciation
Speciation 14
Selection and change in allele frequency 16
Classification
Classification 18
Ecosystems
Ecosystems, ecological terms and ecological techniques 20
Diversity 22
Succession and climax communities 24
Photosynthesis
Photosynthesis 26
Energy transfer in communities
Energy transfer 28
Energy supply 30
Respiration
Respiration 32
Nutrient cycles
Nutrient cycles 34
Human activities and ecosystems
Deforestation 36
End-of-module questions **38**

Module 6: Physiology and the Environment B

Introduction **40**
Water and flowering plants
Transport in plant roots 42
Transport in the xylem 44
Transpiration 46
Homeostasis
Principles of homeostasis and regulation of blood sugar 48
Regulation of body temperature 50
Methods of removing nitrogenous waste 52
Functions of the liver and kidney
Kidney 1 - Urine production 54
Kidney 2 - Water balance 56
Gaseous exchange surfaces
Gaseous exchange surfaces 58
Transport of respiratory gases
Limiting water loss 60
Transport of respiratory gases 62
Digestion and absorption
Digestion and absorption of food 64
Control of digestive secretions 66
Histology of the ileum 68
Metamorphosis and insect diet 69
Neurones, action potentials and synapses
Neurones and action potentials 70
Synaptic transmission 72
Receptors
Receptors 74
Monochromatic and colour vision 76
Control of behaviour
Autonomic nervous system 78
Behaviour 80
End-of-module questions **81**

Module 7: The Human Lifespan HB

Introduction **84**
Sexual reproduction
Gamete formation and fertilisation 85
The developing fetus and maternal system
Implantation and the developing fetus 88
Changes during pregnancy 90
Growth and development
Growth and development 92
Growth and puberty 94
Principal nutrients in the diet 96
Dietary requirements
Dietary requirements 98
Muscles
Muscle structure and the sliding-filament theory of muscle contraction 100
Control of muscle contraction and muscles as effectors 102
Senescence
Senescence 104
End-of-module questions **105**

Appendix: Exam Tips **108**
Answers to quick check questions **111**
Answers to end-of-module questions **118**
Index **121**

Introduction – How to use this revision guide

This revision guide is for the AQA Biology A2 course, Specification A. It is intended for pupils taking either 'Biology' or 'Human Biology'. It is divided into three modules, details of which are given on the next page.

Pupils taking the 'Biology' course should use the pages displaying this symbol ... **B**

Pupils taking 'Human Biology' should follow the pages displaying this symbol ... **HB**

Whichever course you are following, you will be taking tests – either in January or all at the end of the course in the summer.

Each module begins with an **introduction**, which summarises the content.

The content of each module is presented in **blocks**, to help you divide up your study into manageable chunks. Each block is dealt with in several spreads. These do the following;

- summarise the **content**;
- indicate **points to note**;
- include **diagrams** of the sort you might need to label in tests;
- provide **quick check** questions to help you test your understanding. A box in the text, like the one shown here, indicates when you should be able to attempt a particular question.

✓ *Quick check 3*

At the end of each module, there are longer **end-of-module questions** similar in style to those you will encounter in tests. **Answers** to all questions are provided at the end of the book.

You need to understand the **scheme of assessment** for your course. This is summarised on the next page, (But please note that there have been many changes imposed by government in recent times. The details in the book are accurate at the time of publication.) At the end of the book you will find some **exam tips** to help you prepare for the examinable component of the course.

AQA A2 Biology – Assessment

To get a full A level award for AQA Specification A you need an AS (forming 50% of an Advanced Level), followed by the A2 (for the other 50%).

The A2 offered by AQA Specification A offers two alternatives: 'Biology' or 'Human Biology'.

- **Module 5: Inheritance, Evolution and Ecosystems** is common to both courses.
- **Module 6: Physiology and the Environment** is for those taking A2 Biology.
- **Module 7: The Human Lifespan** is for those taking A2 Human Biology.

Note that there are some topics common to both Modules 6 and 7. In this book, these common topics are included in Module 6 only (to avoid repetition). Students taking 'Human Biology' will need to refer to pages marked **HB** in Module 6.

About the tests

To get an A2, you will have to take one exam for Module 5, one exam for Module 6 or 7 and a synoptic exam. The exams for Modules 6 or 7 also have a 5% synoptic content. Synoptic questions are explained in 'Exam Tips' but you will have to know important principles from all the modules you have studied.

What you need to know is given in this book, at the level you need to know it. If you know more than that, it certainly will not harm you but it will not be needed to pass an exam.

The table shows an outline of the AQA Specification A for A2.

<table>
<tr><th colspan="2">A2 Examination</th></tr>
<tr><td colspan="2">Inheritance, Evolution and Ecosystems – Assessment Unit 5
Biology and Human Biology
$1\frac{1}{2}$ hour exam 15% of total A level mark</td></tr>
<tr><td>Either</td><td>Or</td></tr>
<tr><td>Physiology and the Environment – Unit 6 (Biology only)
$1\frac{1}{2}$ hour exam 15% A level
(including 5% synoptic)</td><td>The Human Lifespan – Unit 7 (Human Biology only)
$1\frac{1}{2}$ hour exam 15% A level
(including 5% synoptic)</td></tr>
<tr><td>Unit 8(a)
$1\frac{3}{4}$ hour exam 10% A level
(All synoptic)</td><td>Unit 9(a)
$1\frac{3}{4}$ hour exam 10% A level
(All synoptic)</td></tr>
<tr><td colspan="2">Unit 8(b)/9(b) (Biology and Human Biology)
Centre-assessed Coursework 10% of the total A level marks</td></tr>
</table>

B

HB

Module 5: Inheritance, Evolution and Ecosystems

This module is broken down into 10 topics: Transmission of genetic information, Investigating variation, Selection and speciation, Classification, Ecosystems, Photosynthesis, Energy transfer in communities, Respiration, Nutrient cycles and Human activities and ecosystems.

Transmission of genetic information

- Genetic information is transmitted from generation to generation.
- Meiosis keeps the chromosome number constant from one generation to the next and produces genetic variation.
- Principles of Mendelian inheritance allow prediction of results of genetic crosses.

Investigating variation

- Variation is found between members of a population of a species.
- This can be measured and the data displayed graphically.
- Variation can be due to genetic and environmental factors.

Selection and speciation

- A population has a gene pool, consisting of all the alleles in the population.
- The frequency of alleles can change due to selection.
- This can lead to reproductive isolation of populations and speciation.

Classification

- The basic unit of classification is the species.
- Species belong to one of five Kingdoms; the largest unit of classification.

Ecosystems

- There are ecological terms which have to be used and understood.
- Techniques exist for measuring numbers and distributions of organisms.
- The more hostile an environment, the lower the diversity of organisms.
- A succession takes place on bare soil, leading to a climax community.

Photosynthesis

- Light-dependent and light-independent reactions occur in photosynthesis.
- Light energy is used to produce reduced coenzyme, NADPH and ATP.
- These are used in a reduction reaction that produces sugar molecules.

Energy transfer in communities

- Energy is transferred along food chains and webs.

Respiration

- Many different biological molecules can be used as respiratory substrates.
- These are oxidised by removal of hydrogen, producing reduced coenzymes.
- Reduced coenzymes provide the energy for oxidative phosphorylation, which produces most of the ATP in aerobic respiration.
- ATP is used as the immediate source of energy for other processes.

Nutrient cycles

- Carbon enters biological molecules via photosynthesis and leaves as carbon dioxide from respiration.
- Nitrogen is recycled by microorganisms which decompose dead organisms.

Human activities and ecosystems

- Deforestation reduces diversity and disrupts nutrient cycles.

B
HB

Meiosis (and inheritance)

The first twelve pages of this module concern topics requiring understanding of some commonly used genetic terms which are dealt with below.

Genotype

A gene is a specific length of DNA carrying coded information for producing a particular protein/polypeptide. It is found at a specific place/**locus** on the DNA molecule in a chromosome.

- **Genotype** – the **genes an organism has and the alleles of those genes.**
- **Phenotype** – the characteristics of an organism, including the enzymes and substances it produces. It depends on expression of the genotype and **interactions between the genotype and environment.**

- **Diploid** organisms inherit **two copies of each gene and chromosome**; one from each parent (maternal and paternal).
- **Haploid** cells or organisms have **one copy of each gene and chromosome.**
- A gene can exist in different forms, called **alleles.**
- **Homozygous** – having two identical alleles of a gene.
- **Heterozygous** – having two different alleles of a gene.
- **Dominant** alleles – expressed when homozygous or heterozygous.
- **Recessive** alleles – only expressed in the homozygous state.

Dominant and recessive have nothing to do with 'better' or 'worse' – or more or less common. Huntington's chorea is a fatal condition caused by a dominant allele. Blood type O is the commonest blood type in the UK and is homozygous recessive.

Quick check 1

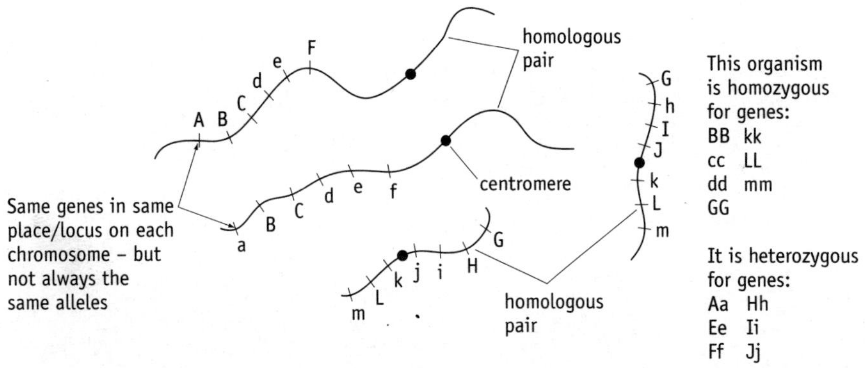

homologous pair

centromere

Same genes in same place/locus on each chromosome – but not always the same alleles

homologous pair

This organism is homozygous for genes:
BB kk
cc LL
dd mm
GG

It is heterozygous for genes:
Aa Hh
Ee Ii
Ff Jj

Chromosomes of a diploid cell with 4 chromosomes – 2 pairs

Meiosis

Sexual reproduction involves the production of **haploid** male and female **gametes,** which fuse at **fertilisation.** If gametes carried the diploid number of chromosomes or genes, the chromosome number would double at fertilisation. Meiosis at **some point** in the life of an organism **keeps the chromosome (and gene) number constant from one generation to the next.**

Ⓢ A gene is a sequence of bases on DNA, carrying information that can be transcribed into mRNA and translated into polypeptides at ribosomes.
New alleles arise by mutation.

Ⓢ In some organisms the adults are haploid and produce gametes by mitosis. – meiosis occurs after fertilisation.

- **Meiosis** produces **haploid cells** with **one copy of each** gene – **it reduces the chromosome number**.

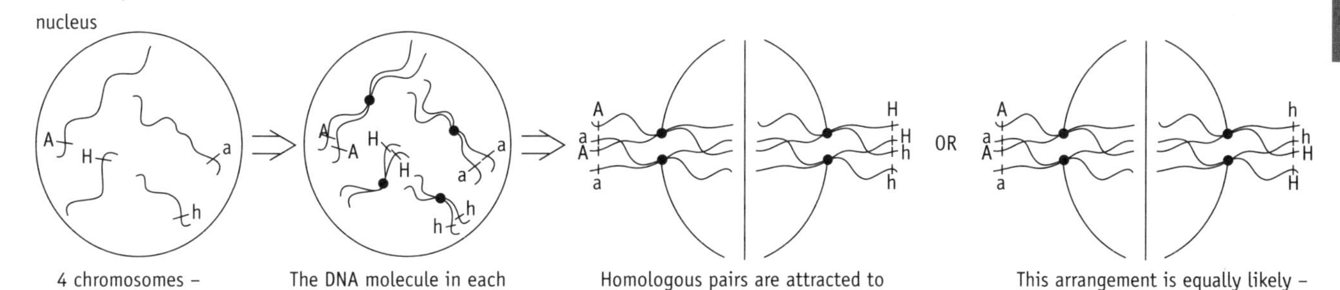

Independent assortment in meiosis

- At the start of meiosis, the DNA molecule in each chromosome replicates, forming two identical sister **chromatids**, held together at the **centromere**.
- Chromosomes in diploid organisms form **homologous** pairs, one of maternal and one of paternal origin, carrying the same genes but not always the same alleles.
- In meiosis, homologous chromosomes come together and their chromatids wind round each other, forming **chiasmata**.
- Pairs of homologous chromosomes, held together at chiasmata, are called **bivalents.**
- Pieces of chromatid are sometimes exchanged between homologous chromosomes at a chiasma.
- This is **crossing over** which produces **new combinations of alleles**.

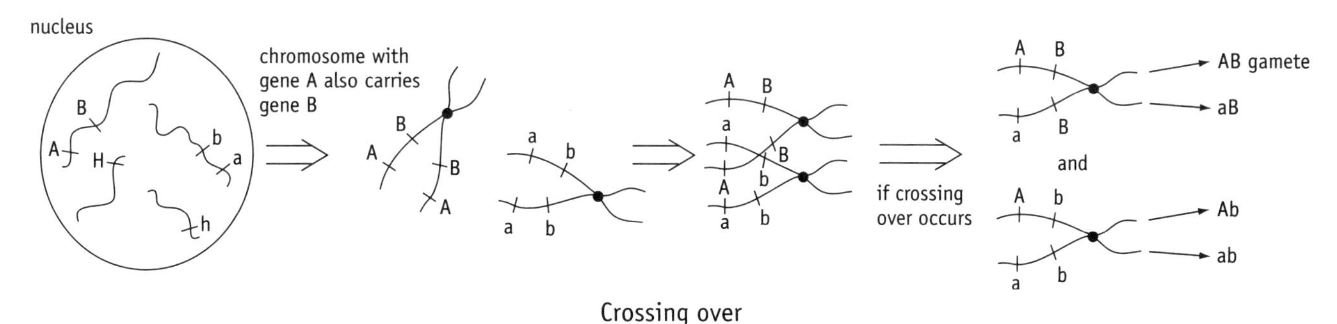

Crossing over

- Bivalents line up on the equator of the meiotic spindle and one of each pair moves to each pole of the spindle, **halving the chromosome number**.
- The cell divides, producing **haploid** cells.
- Bivalents can line up on the spindle to give any combination of maternal and paternal origin chromosomes after division – this is the **independent assortment of homologous chromosomes.**
- This produces haploid sets of chromosomes containing **new combinations of maternal and paternal origin chromosomes and alleles** and genes.
- The haploid cells' chromosomes still consist of two chromatids, which are separated during a second cell division, to give four haploid cells.

Do not try to learn the names and details of all of the stages of the first meiotic division – it is not required and no marks are awarded for knowing them.

Quick check 2, 3

Chromatids are held together at a centromere. When they separate during cell division, each chromatid becomes a chromosome.

Quick check questions

1. Explain what is meant by each of the following; (**a**) genotype, (**b**) phenotype.
2. Explain why meiosis is necessary at some stage in the life of an organism that reproduces sexually.
3. Look at the homologous chromosomes in the figure at the top of the page and explain how a gamete with the genotype AbH could be produced.

B

HB

Principles of Mendelian inheritance

Monohybrid inheritance

Monohybrid crosses involve **alleles of a single gene.**

- **Recessive** alleles can be carried – **carriers** are **heterozygous** for the allele and unaffected.
- Recessive alleles can cause genetic disorders, if a person is homozygous, e.g. **cystic fibrosis.**
- If two carriers have children, each child has a 25% (1 in 4) chance of inheriting two copies of the cystic fibrosis allele and being affected.
- Questions might involve two generations of crosses, the F_1 and F_2. The example shown below involves tall and short pea plants.
- Homozygous (pure-breeding) tall plants were crossed with homozygous (pure-breeding) short plants and the F_1 offspring were grown.
- All the F_1 were heterozygous and tall; showing that tall is dominant.
- The F_1 were crossed with each other, producing the F_2 generation.
- The $\mathbf{F_2}$ contained tall and short plants in a **3:1 ratio** – a **phenotypic ratio.**
- F_2 **genotypes** 1 homozygous tall: 2 heterozygous tall: 1 homozygous short.

> In genetics questions, set out details clearly in diagrams. It is important to read information given in questions; it may help you to find the genotypes of parents in crosses.

> Use a capital letter for a dominant allele, e.g. T, and the small case of the **same letter** for the recessive, t. Do **not** use a different letter!

Parental genotypes — Tall pea plants X Short pea plants

TT — tt

Gametes (T) (t)

F_1 all Tt

Crossing the F_1 plants with each other

Tt X Tt

Gametes (T)(t) (T)(t)

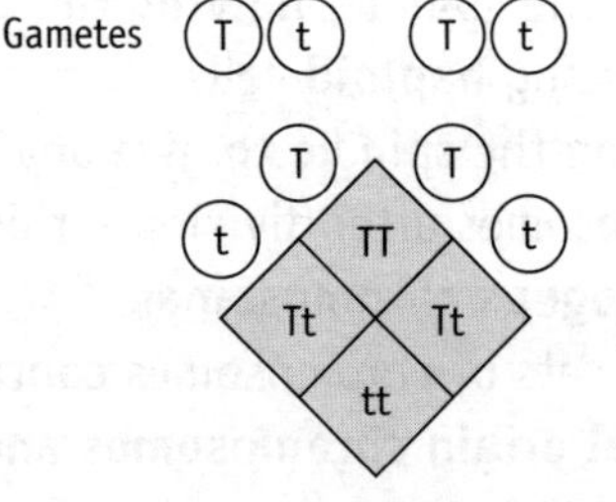

F_2 genotypes TT , Tt , Tt , tt

F_2 phenotypes 3 tall : 1 short

A monohybrid cross

Multiple alleles and codominance

- A diploid organism has **two copies of each gene.**
- Many genes have more than two alleles; these are **multiple alleles.**
- In an organism homozygous for a gene, both alleles are the same.
- A heterozygous organism has two different alleles of the gene – and that is all it can have, no matter how many alleles the gene has.
- In some cases one allele is dominant over the other (which is recessive) and is expressed – shown in the phenotype
- Some alleles show **codominance** – both are expressed in the phenotype of heterozygous individuals.
- A dominant allele usually produces a functional protein and a recessive allele does not.

> Read questions carefully. If they ask for the genotypes of offspring, that will be TT and Tt and tt, for e.g. If they ask for phenotypes, the answer will be tall or short.

> A 3:1 ratio is the same as 1 in 4, or 75% to 25%.

- Codominant alleles both produce functional protein; example, ABO blood groups in humans. See the first figure below.

✓ Quick check 2

Sex-linked characteristics

Genes on the X chromosome are sex-linked – you should assume that the Y chromosome has no functional genes on it.

- Women can be homozygous or heterozygous for sex-linked genes.
- Men inherit **one** copy of a sex-linked gene, on the X chromosome from their mother, which is **always** expressed in their phenotype.
- **Haemophilia** is a recessive sex-linked disorder in humans.
- A boy who inherits this allele from his mother will suffer from haemophilia.
- This is why haemophilia is much more common in men than women.

The ABO bloodgroups are determined by 3 alleles of one gene – I^A, I^B and I^O. I^A and I^B are codominant and both are dominant over I^O (a recessive)

Look at this example – both parents carry I^O

Parental genotypes: Father – type A I^A I^O; Mother – type B I^B I^O

gametes: (I^A) (I^O) (I^B) (I^O)

Father's gametes \ Mother's gametes	I^B	I^O
I^A	$I^A I^B$	$I^A I^O$
I^O	$I^B I^O$	$I^O I^O$

Possible genotypes of children: $I^A I^B$, $I^A I^O$, $I^B I^O$, $I^O I^O$

Possible phenotypes of children – blood group: AB, A, B, O

Inheritance of ABO blood types – multiple alleles – dominant, recessive and co-dominant alleles.

✓ Quick check 3

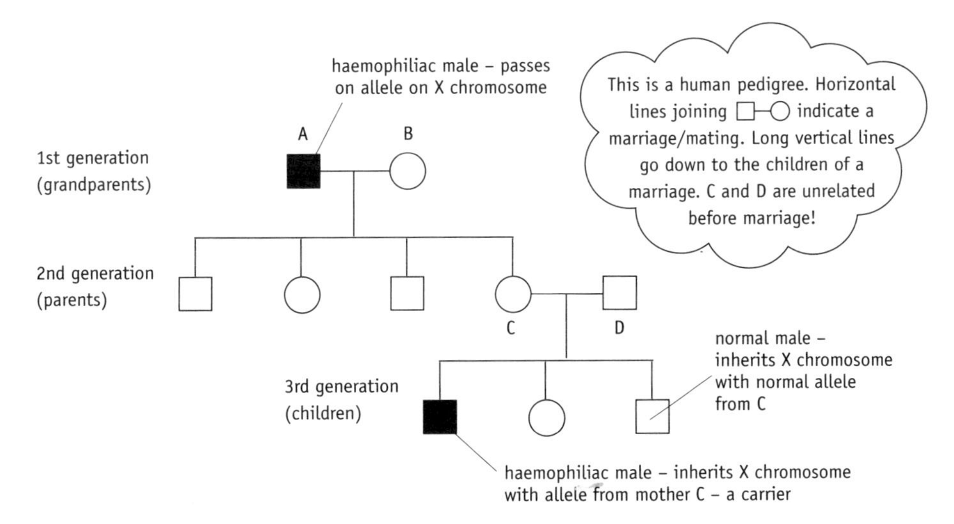

Sex-linkage in inheritance of haemophilia

It is never good enough to say that a gene is sex-linked because more of one sex show it than another. Questions usually give **evidence** that a gene is on the X chromosome – you have to find it! Look at where the X chromosomes are inherited from.

Quick check questions

1. A pure-breeding round pea plant was crossed with a pure-breeding wrinkled pea plant. All of the F_1 had round peas. Explain the offspring you would expect in the F_2.
2. A man has blood type A and is heterozygous. His wife is blood type B and heterozygous. Explain the possible genotypes of their children.
3. A couple whose first child has haemophilia consult a genetic counsellor about their chances of having another affected child. Suggest the sort of advice they could be given.

B

HB

Dihybrid crosses and chi-squared

These crosses involve **alleles of two genes.**

The example looks at inheritance in pea plants of characteristics of tall or short and yellow or green pea colour.

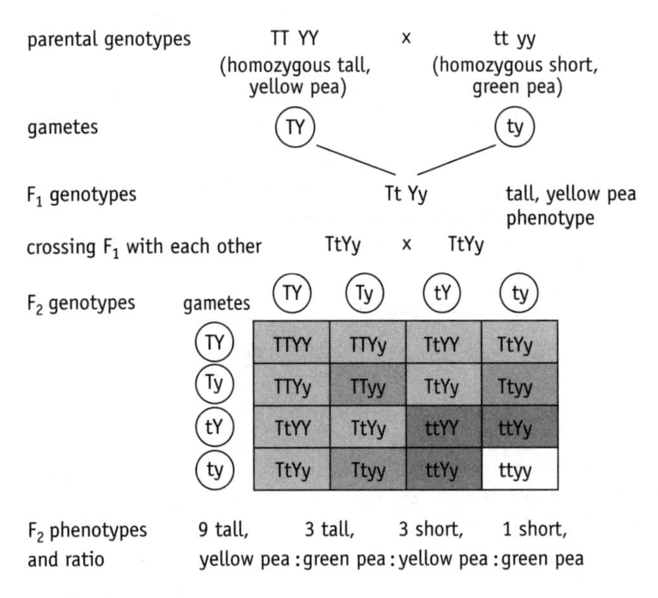

Dihybrid cross

- Homozygous (pure-breeding) tall, yellow pea plants were crossed with homozygous (pure-breeding) short, green pea plants.
- All F_1 were tall with yellow peas, showing that tall and yellow are dominant.
- The F_1 were crossed with each other (selfed), to produce an F_2 generation.
- The F_2 contained a ratio of 9 tall, yellow pea: 3 tall, green pea: 3 short, yellow pea: 1 short, green pea.
- This is a typical **F_2 ratio of 9:3:3:1**.
- Note how the **Punnett square** is constructed to show the gametes produced.
- Each gamete contains **one** allele of each gene.
- You should be aware that the F_2 ratio is the result of **independent assortment** of homologous chromosomes in **meiosis** leading to gamete production.

In using Punnett squares, make sure that you indicate which squares represent the gametes (and which parent they are from).

You may well get an example in a question involving genes which you are not familiar with. The F_1 results usually give evidence of which alleles of genes are dominant and which recessive. This evidence should be given in your answer.

✓ Quick check 1

Expected ratios, observed ratios and the chi-squared test

On page 6 the F_2 ratio in the monohybrid cross is 3:1 and on this page the F_2 ratio in the dihybrid cross is 9:3:3:1. These are **expected ratios, calculated** from genotypes of the parental generation and assuming independent assortment, no sex linkage, or codominance. In real crosses the offspring produced depends on chance fusions of gametes (fertilisation), leading **to observed ratios** – from **results actually obtained.** There are differences between expected and observed ratios and the question is, are these differences due to chance alone, or are they **statistically**

B

HB

significant? If they are, then the assumptions used to predict the expected ratio might be wrong, for example, the gene(s) might be sex linked.

Chi-squared (χ^2) is used to decide if differences between sets of results/data are significant.

$$\chi^2 = \Sigma \frac{(O - E)^2}{E}$$

- **O** are **observed results** – measured or recorded results.
- **E** are **expected results** – calculated from observed results.
- **Null hypothesis** – there are **no significant differences between sets of data.**

Example: In a monohybrid cross between short and tall pea plants, the F_2 consisted of 175 tall and 46 short plants. Is this significantly different from a 3:1 ratio? The null hypothesis is that it is not.

- **The expected results** are calculated by adding the observed results together and dividing by 4 to find expected short plants.

	Phenotypes of F_2	
	Tall	**Short**
Observed (O)	175	46
Expected (E)	165.75	55.25
O - E	9.25	–9.25
$(O - E)^2$	85.56	85.56
$(O - E)^2 \div E$	0.52	1.55
Sum of $\Sigma(O - E)^2 \div E$	chi-squared = 2.07	

$$\chi^2 = \Sigma \frac{(O - E)^2}{E} = 2.07$$

- There are tables of values for χ^2, for different degrees of freedom; in this example 1 – one less than the number of classes of data.
- A **probability value** of χ^2 **less than 0.05**, means a **significant difference** between observed and expected results.
- The χ^2 value of 2.07, for one degree of freedom, comes between probability values of 0.20 and 0.10.
- You would expect this amount of difference between observed and expected results between 10 and 20% of the time by chance alone.
- There is no significant difference from the expected results and the null hypothesis is accepted.

✓ Quick check 2

? *Quick check questions*

1. Pea plants homozygous for purple petals and inflated pods were crossed with pea plants homozygous for white petals and constricted pods. All the F_1 plants produced purple petals and inflated pods. Explain the results you would expect if the F_1 were crossed with each other.
2. **(a)** Explain what is meant by expected and observed results and how they are obtained. **(b)** Explain what is meant by a significant difference between sets of data.

B

HB

Investigating variation

A species exists as populations, groups of organisms of the same species, living in the same habitat and able to interbreed.

There is **variation** in any population – the individuals are not all identical.

Types of variation

Variation can be continuous or discontinuous.

Continuous variation

- **Characteristics having a range between two extremes.**
- E.g., height and mass of humans show a range from smallest to greatest.
- These characteristics can be **measured**, in units such as length or mass – they are **quantitative.**
- There are no separate categories or types, only differences of degree.
- Data can be presented in **line graphs** or **histograms.**

Ⓢ A population is all the organisms of one species in a habitat. There is intraspecific competition between members of a population. Where there is variation, some phenotypes will be better adapted to the environment and more likely to survive to reproduce.

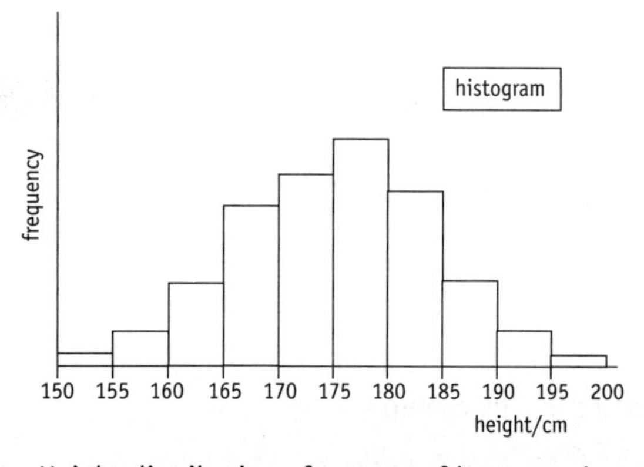

Height distribution of a group of human males

Discontinuous variation

- **Characteristics showing separate/discrete categories or classes,** e.g. human ABO blood types – everyone falls into one of the types.
- There are no intermediate types.
- These differences can not be measured in units – they are **qualitative.**
- Data can be presented in **bar charts** or **pie charts.**

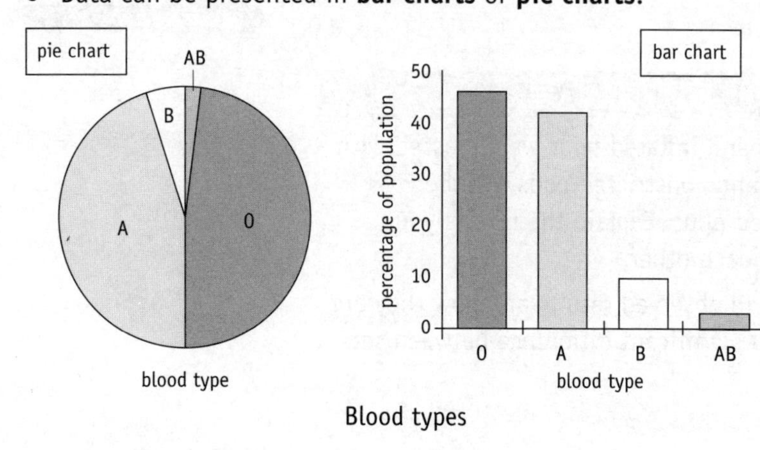

Blood types

Environmental factors can affect characteristics showing discontinuous variation. For example, pea plants can be genetically tall or short but slightly different environments for each plant can produce a range of heights for both tall and short plants.

✓ Quick check 1

B

HB

Measuring variation in populations

We often want to measure or record characteristics of individuals in a population. For example, what is the mean height of men, or what is the commonest blood type in Britain?

- With large populations it isn't easy/possible to measure or record data from every individual.
- Data is recorded from **many (20+) samples**, because one sample is unlikely to accurately represent the whole population.
- There will be **differences** between samples due to **chance** differences in **genes, alleles or the environment of a sample**.
- There have to be **enough samples**, of a **large enough size** to be **representative** of the whole population and overcome chance differences.
- **Sampling should be random**, to avoid any prejudice (conscious or unconscious) on the part of the person collecting the data – they might pick certain types to be sampled.

Random number tables/generators might be used. For example, an area to be sampled might be divided into numbered grids and then random number tables used to select the grids to be sampled.

Standard deviation and standard error

Standard deviation measures spread of data about the mean.

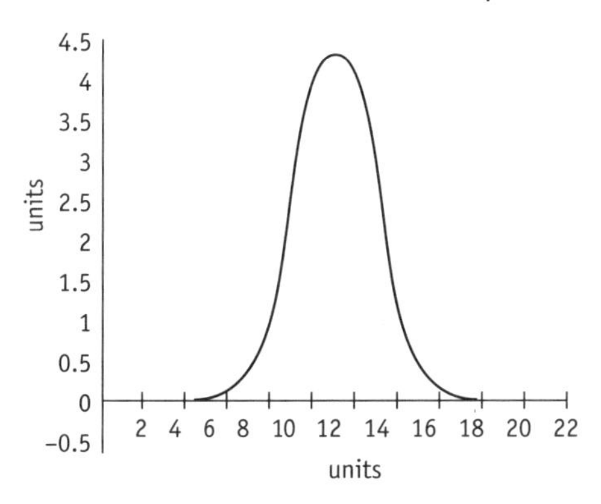

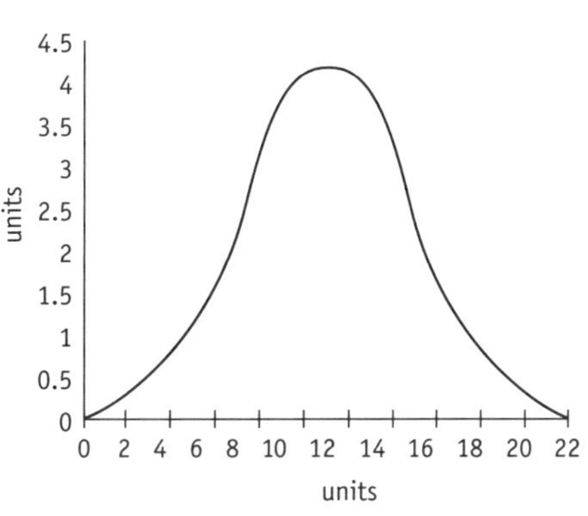

Standard deviation

- These sets of data have the **same mean**.
- The data shows **normal distribution** about the mean value – there is a bell-shaped and even distribution of values above and below the mean.
- The **smaller** the standard deviation, the **less the variation** between individuals for the characteristic measured.
- The **greater** the standard variation, the **greater the variation**.

Standard error is the **standard deviation of the mean**.

- If individuals in a number of samples are measured, then each sample will have its own mean.
- These means will usually be slightly different from each other – reflecting chance differences in samples – giving a range of values for sample means.
- Standard error is a measure of how much the value of a sample mean is likely to vary.
- The greater the standard error, the greater the variation of the mean.

✓ Quick check 2

? *Quick check questions*

1. Explain how you would present each of the following types of data in a graphical form: (**a**) human masses, (**b**) human eye colour.
2. Explain the difference between the standard deviation and standard error of a set of data.

B

HB

Causes of variation

There are two main causes of variation, **genetic factors** and **environmental factors.** The greatest similarities come when individuals share identical (or very similar) genotypes (identical twins and clones) and are raised in the same environment.

In discontinuous variation:

- there is a **strong genetic factor**,
- often due to **alleles of a single gene**,
- **environmental factors play a small/no part**, e.g. ABO blood groups.

Sometimes, environmental factors affect discontinuous variation, e.g. tall and short pea plants.

- Genetically tall pea plants grown under different environmental conditions, produce a range of heights but are taller as a group than genetically short plants.

In continuous variation:

- there is a **strong environmental factor**(s),
- (where genetic factors are involved,) **many genes** are involved – **polygenic**,
- environmental factors play a part, e.g. many genes contribute to height in humans but a lack of food, or particular food types, will lead to stunted growth, regardless of the person's genotype.

✓ Quick check 1

Sources of genetic variation

When genetic factors are involved in variation, you need to be aware of what produces genetic variation – leading to different genotypes.

Mutation

This is the ultimate source of genetic variation, producing **new alleles** of genes (and sometimes new genes).

- **Substitution** mutations involve substitution of one base in a DNA sequence by another base, **altering one triplet** of bases.

Ⓢ The genetic code is degenerate. A substitution may produce a different triplet but one that codes for the same amino acid!

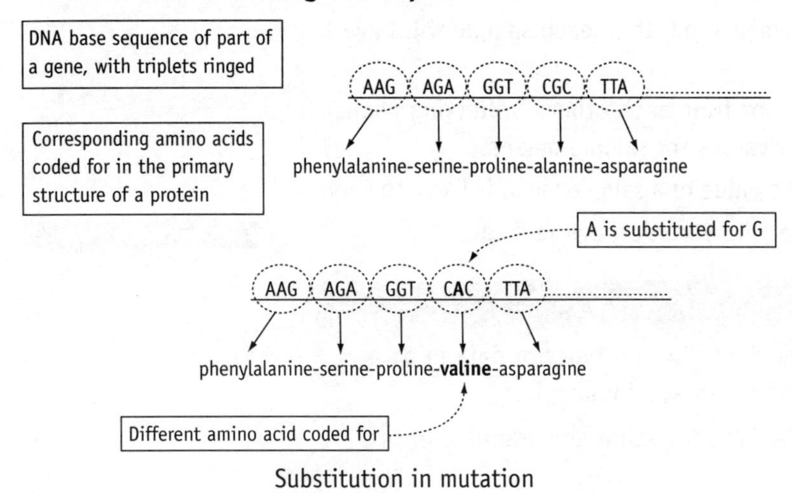

Substitution in mutation

B

HB

- The new triplet usually codes for **one different amino acid** in the primary structure of the protein coded for by the gene.
- The effect this has depends on the effect it has on the tertiary structure of the protein produced.

Deletion

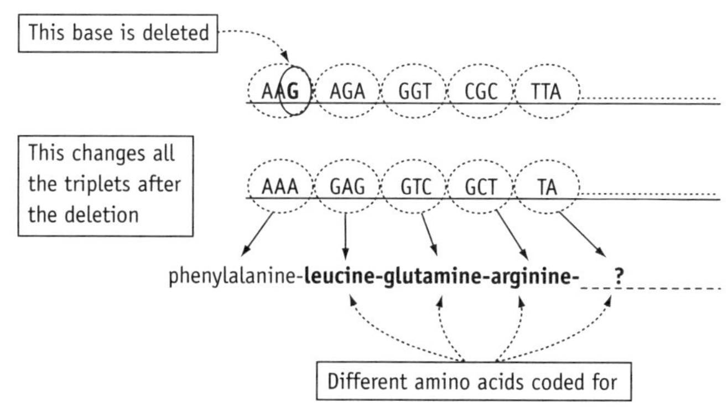

Deletion in mutation

Sickle-cell anaemia is caused by a difference in one amino acid in haemoglobin.

- Deletions are usually harmful, because large numbers of amino acids are changed – producing a protein with a completely different tertiary structure.
- Sometimes mutations can produce new genes, or combinations of chromosomes.

Quick check 2

Meiosis produces genetic variation through;

- **crossing over** at chiasmata – producing new combinations of maternal and paternal origin alleles on homologous chromosomes,
- **independent assortment** – producing new combinations of maternal and paternal origin chromosomes and their alleles.
- Both produce genetic variation in gametes involved in sexual reproduction.

Random fusion of gametes at fertilisation also brings together new combinations of alleles.

Meiosis does not cause mutations! There are some mutations associated with mistakes in meiosis but not ones that concern this course.

Quick check 3

Quick check questions

1. Explain whether each of the following is an example of continuous, or discontinuous variation: human IQ, haemophilia, length of hair.
2. A pair of identical twins were raised apart and so were a pair of non-identical twins. Explain which of the following characteristics you would expect to vary more in the non-identical twins: height, mass, ABO blood group.
3. Explain how meiosis increases genetic variation.

B

HB

Hardy-Weinberg principle

The gene pool is **all the genes and their alleles** in a population. Organisms in a population have the **same genes** but **different combinations of alleles.**

- Alleles pass from one generation to the next during reproduction.
- **Sexual reproduction** produces **new combinations of alleles.**

Allele frequency is the frequency of an allele in a population.

- Each organism (diploid organisms) has two copies of each gene and is either homozygous or heterozygous for the alleles of that gene.

Example: **Albinos** can not make melanin (a pigment) that colours skin, hair and eyes.

- Humans making melanin are homozygous AA or heterozygous Aa and albinos are homozygous aa.
- The **Hardy-Weinberg equation** allows calculation of probabilities genotypes.
- Let p represent the probability of allele A and q the probability of a:

$$p^2 + 2pq + q^2 = 1$$

- One in 40,000 is an albino, a probability q^2 = **0.000025** <0.0025%>
- The square-root of this gives a value for **q of 0.005.**
- The probability of Gene A is 1, i.e. 100% (all organisms have the gene) so, **p + q = 1.**
- If **q = 0.005, then p = 1 – 0.005 = 0.995.**
- The probability of homozygous AA is $(0.995)^2$ = 0.990025 (99.0025%)
- The probability of heterozygous Aa is 2 x 0.005 x 0.995 = 0.00995 (0.995%)

Ⓢ The DNA of the recessive allele codes for a non-functional enzyme and this blocks the metabolic pathway that synthesises melanin. Heterozygous people produce enough functional enzyme to be able to make melanin.

Figures are usually given to three significant figures but in this example more are needed – to show that the numbers add up to 1!

The **Hardy-Weinberg equation** applies where:

- the population is diploid and sexually reproducing,
- particular alleles do not directly affect differential mortality, reproduction or mate selection, or migration,
- the population size is large,
- there are no mutations.

✓ *Quick check 1*

Speciation

A **species is a group of organisms that can interbreed to produce fertile/viable offspring.** New species arise from existing species by the process of **speciation.**

- **Evolution is a change in the frequency of alleles in a population.**
- Organisms in a population which are better adapted are more likely to survive, reproduce and **pass on their combination of alleles/genes to the next generation –** there are **differential survival rates.**
- This **changes the frequencies of alleles and phenotypes** in a population – more advantageous ones increase in frequency, less advantageous decrease.

The definition of evolution makes no mention of creating new species, because it does not have to do this – in fact, it usually doesn't!

B

HB

- **Natural selection** is the process by which an environmental factor(s) affects the survival rates of different phenotypes in a population.
- Each **population** of a species lives in a different (but usually similar) environment.
- Different selection pressures cause each population to evolve differently from all the others – and adapt to its environment
- **Emigration and immigration** usually move alleles between populations, preventing big differences from developing.

Ⓢ In a new environment, a population may find niches that are unoccupied by other species and show little interspecific competition for some important factor – such as food. This can lead to adaptive variation – where the population evolves adaptations that allow it to occupy new niches.

Allopatric speciation

- Sometimes a population becomes **isolated**, usually due to **geographical isolation**; e.g; populations on islands, different continents, or either side of mountain ranges.
- An isolated population adapts to its environment through natural selection.
- Changes in genotypes and phenotypes may lead to **reproductive isolation**.
- They can not breed with members of other populations of the original species to produce viable offspring – they are now a **different species**.

Woodpecker finch
bores holes in wood to find insects

Warbler finch
catches insects on the wing

Cactus finch
feeds on nectar from cactus flowers

Ground finch
typical finch beak for crushing seeds

Speciation – Darwin's finches found on the Galapagos Islands

Sympatric speciation

This occurs when reproductive isolation evolves between members of the same population, living in the same area.

- Some members of a population become able to exploit a new/different niche.
- A mutation could make this possible, or natural selection against members of a population temporarily geographically isolated from the rest.
- Reproductive isolation then occurs, perhaps by another mutation.

You would test to see if two similar organisms belong to the same species by mating them. If they produce viable/fertile offspring, then they belong to the same species.

✓ *Quick check 2, 3*

? *Quick check questions*

1 Height in pea plants can be determined by two alleles of one gene, tall and short. In a population of pea plants, 9% were homozygous for the recessive short allele. Calculate (**a**) the frequency of the tall allele, (**b**) the frequency of heterozygous pea plants in the population.

2 Explain why differential survival rates lead to evolution of a population.

3 A new species of beetle was found on an island. It was clearly closely related to a species found on the mainland but research showed that it was a different species. (**a**) Suggest how you could prove that the beetle was a new species. (**b**) Explain how the new species evolved.

B

HB

Selection and change in allele frequency

In **natural selection, environmental factors affect the survival rates of different phenotypes** in a population. Selection can be stabilising, directional or disruptive.

Ⓢ A population lives in an ecosystem and forms part of a community. Organisms of the population will form part of food chains and webs. Anything that affects the stability of the population will affect other populations of other species.

Stabilising selection

In a **stable environment**, natural selection over time usually favours 'average' members of a population; those 'on average' best adapted to that environment.

- Organisms in each generation with extreme forms of characteristics, or mutations, are selected against.

Example: Birth weight in humans

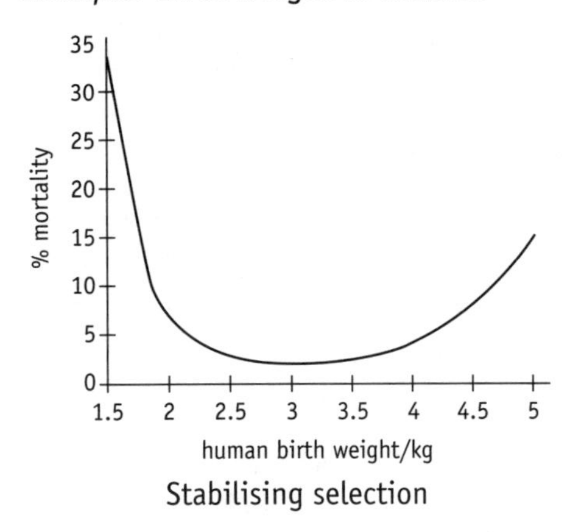

Stabilising selection

- The heaviest and lightest babies have the highest mortality and are less likely to survive to reproduce and pass on their alleles.

✓ *Quick check 1*

Directional selection

An **environmental change** may produce new selection pressures that favour organisms with an extreme form of a characteristic.

Pesticide resistance – *Example:* Warfarin – a poison used to kill rats.

- When warfarin was introduced, some populations **already contained rats** with a **chance mutation** that gave them resistance to the poison
- Without warfarin, stabilising selection favours normal rats – resistant rats are selected against, because they need a lot of vitamin K in their diet.
- Warfarin was a new environmental factor that killed normal rats.
- A few resistant rats survived, reproduced and passed on the resistance gene.
- They produced a new population of resistant rats.

Antibiotic resistance – *Example:* Penicillin resistance

- When penicillin was introduced, some populations **already contained bacteria** with a **chance mutation** that gave them resistance to the antibiotic.
- These bacteria are usually selected against, because they waste energy making unnecessary enzymes.
- Penicillin killed normal bacteria, leaving resistant bacteria to form a new population.

The commonest mistake is thinking that organisms develop a characteristic **because of** a change in the environment, e.g. that introducing a disease forces a mutation to take place.

In most cases when people take an antibiotic, the infecting bacteria are killed and the person gets better. This means that there weren't any resistant bacteria in the infecting population.

B

HB

Populations do not decide to adapt, or mutate, after an environmental change. The mutation, or combination of alleles giving resistance, have to already be there by chance, otherwise the population may become extinct.

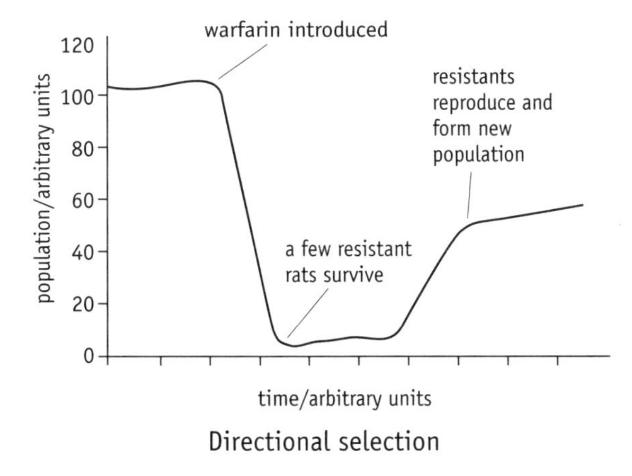

Directional selection

Quick check 2

Ⓢ Genes for antibiotic resistance are often used as 'marker genes' to identify genetically engineered organisms.

Disruptive selection

An **environmental change** may produce selection pressures that favour two extremes of a characteristic.

Sickle-cell anaemia

- People homozygous for this recessive allele usually die before reproducing.
- Their red blood cells contain abnormal haemoglobin which makes them become sickle-shaped and stick in capillaries.
- People heterozygous for the allele should be at a disadvantage, because their red cells can sickle during exercise – the allele should be selected against and rare.
- Its **frequency is high** in parts of the world where **malaria** is common – in some populations over 20% carry the allele (as heterozygotes).
- People **heterozygous** for **sickle-cell** anaemia are **more resistant to malaria** than people homozygous for the normal allele.
- Where malaria is found, people heterozygous for sickle-cell have an advantage and are likely to survive, reproduce and pass on the allele.
- People without the allele also have an advantage, because their red cells behave normally.
- This produces populations with an equilibrium for numbers of people heterozygous for sickle-cell and non-carriers (balanced polymorphism).

Ⓢ Farming practices often change the environment of populations of other species. Fertilisers, effluent and pesticides all apply new selection pressures to populations of other species.

Ⓢ New sickle-cell alleles are produced by chance mutations from time to time.

Quick check questions

1. Explain why some members of the same population are more likely to survive and reproduce than others.
2. An exam candidate wrote that, "The rats became resistant to warfarin, so that they could survive." Explain why this answer is wrong.
3. It might seem that populations in areas with malaria should evolve so that everyone is heterozygous for sickle-cell anaemia. Explain why this does not happen.

B
HB

Classification

The science of classification is known as **taxonomy**. Part of this science is **nomenclature** – the naming of organisms.

The definition of evolution makes no mention of creating new species, because it does not have to do this – in fact, it usually doesn't!

Species

Members of the same species:

- are able to breed together and produce fertile offspring,
- have very similar phenotypes – show very similar variation – in terms of structure, biochemistry and behaviour,
- have a common evolutionary ancestry.

Classification

- The **binomial system** is used for scientific naming of species.
- Each name consists of **two** words, the **generic** name and the **specific** name.
- The generic name starts with a **capital** letter and the specific name starts with a **small case** letter.
- The name is either **underlined, or written in italics**.
- The words used in the name are forms of Latin/Greek.

In classification, organisms are put into **taxa**. A taxon contains organisms which share some basic feature. The different levels of taxon are; **kingdom, phylum, class, order, family, genus, species**. These taxa make up a **hierarchy**.

- The **species** is the **smallest** taxon; it contains only one type of organism.
- A **genus contains** one or more species.
- A **family contains** one or more genera.
- An **order contains** one or more families.
- A **class contains** one or more orders.
- A **phylum contains** one or more classes.
- A **kingdom** is the **largest** taxon and **contains** one or more phyla.

A kingdom is a **composite group**, made up of phyla and a phylum is a composite of classes, and so on until the species is reached – which is not a composite group. The hierarchy means that any organism will have a unique classification. It fits into a certain set of taxa, because there are **no overlaps** between taxa.

One hierarchy is the complete classification of one species – from species to kingdom.

Human classification:

kingdom	Animalia
phylum	Chordata
class	Mammalia
order	Primates
family	Homidae
genus	Homo
species	Homo sapiens

Phylogenetic classification

Phylogenetic classification reflects evolutionary links between organisms and their evolutionary history.

- Species in the same genus share a common ancestral species, from which they evolved.
- Genera in a family share a common ancestor but further back in evolutionary history than the species in a genus.
- Families in an order also share an ancestor but even further back in evolutionary history.

This type of relationship applies all the way up the hierarchy, to the kingdom.

✓ Quick check 1, 2

The five kingdoms

Modern classifications recognise five kingdoms; Kingdom Prokaryotae, Kingdom Protoctista, Kingdom Fungi, Kingdom Plantae and Kingdom Animalia.

The five kingdoms

Kingdom	Distinguishing features
Kingdom Prokaryotae	Bacteria and blue-green bacteria – microscopic, prokaryotic cells.
Kingdom Protoctista	Organisms with eukaryotic cells that are not in the Kingdom Fungi, Plantae or Animalia – often called Protozoa and Algae – unicellular, filamentous, colonial or macroscopic (e.g. seaweeds).
Kingdom Fungi	Eukaryotic with cell walls made of chitin and no cilia or flagella – some unicellular, others consist of thread-like hyphae which form a mycelium – can not photosynthesise and are saprotrophic or parasitic.
Kingdom Plantae	Eukaryotic with cell walls made of cellulose, large vacuoles and chloroplasts (certain cells) – multicellular and photosynthesising – adapted for life on land (most) – growth restricted to meristems (layers/patches of dividing cells).
Kingdom Animalia	Eukaryotic with no cell walls – multicellular – obtain nourishment by feeding, can not photosynthesise – nervous system – growth throughout tissues (no meristems).

Ⓢ You dealt with the differences in cell organelles in prokaryotic and eukaryotic cells in AS Module 1. The main differences are the much smaller size of prokaryotic cells and their lack of membrane bound organelles.

Viruses

Viruses consist of nucleic acid and a protein coat – they have no cellular structure. They are not thought of as being alive in the usual sense and are not included in most biological classifications.

✓ Quick check 3

? Quick check questions

1 Explain what is wrong with the following statement, "The correct name for humans is Homo Sapiens."

2 Recent discoveries in genetics have shown that humans share 98% of their genetic material with chimpanzees. Suggest how phylogenetic classification can explain this.

3 Describe the differences between a plant and a fungus.

B

HB

Ecosystems, ecological terms and ecological techniques

An **ecosystem** is a **community** of populations of species and their abiotic environment.

Ecological terms

Common terms	
Term	**Definition**
population	All the organisms of one species living in a habitat and interbreeding.
habitat	Where an organism lives, including its abiotic and biotic environment.
abiotic factors	Physical conditions; e.g. temperature, pH, rock type, soil particles, wind.
biotic factors	The influences due to other organisms
community	Consists of populations of species in the same habitat – often named after the dominant plant, e.g. oak wood.
niche	Where an organism is found in the habitat and its role in its community – described in terms of the range of abiotic factors it needs and feeding.

- A community has **energy** flowing through its **food web**.
- **Feeding relationships are dynamic** – energy transfer changes with time with changes in numbers of organisms in each trophic level.

You must learn the definitions in the table!

Ecological techniques

Line transects and frame quadrats

- **Line transects** study changes in vegetation types from one area to another.
- A tape measure/string is stretched across areas being studied.
- Plant species under the line are recorded at regular intervals along the tape/string; perhaps using a frame quadrat.
- A **limitation** is that one transect may not cross typical areas, so they are **repeated several times**, along random lines and the results averaged.
- A **frame quadrat** is of known area (usually 0.25 m^2 or 1 m^2), with wires/strings stretched across it.

The ecosystem described in a question might not be one you have studied but the methods you have used in your studies will still apply.

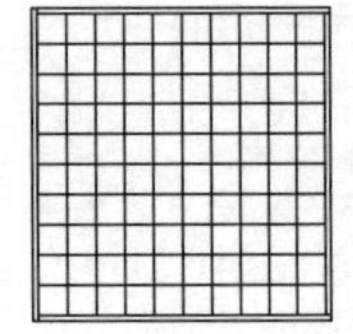

A quadrat

- **Percentage cover** – by counting the quadrat squares covered by a plant.

B

HB

- **Limitations** of the frame quadrat:
  - estimating the number of squares covering each species,
  - the small area of the quadrat, compared with the area being studied.

Frame quadrats can be:

- placed next to a line transect at regular intervals,
- used to find percentage cover of a plant in an area, by throwing it at random – to avoid picking 'interesting'areas.
- placed at points on a grid chosen using random number tables.
- Results are taken from many (20+) quadrats and the results averaged.

Abiotic factors are important because,

- physiological adaptations of organisms only allow them to live in a certain range of e.g. pH, light and temperature – it is part of what defines their niche.

Many candidates give poor answers to 'methods' questions, because they fail to give details – e.g. what exactly is recorded, units, how often, the need for many results/repetitions and averages.

✓ Quick check 1

Mark–release–recapture techniques

It is often difficult to count all the animals in a population, because most animals move around. It is possible to get an estimate of population size using capture-recapture methods.

- A large number of animals are caught (20+), to try and make sure that it is a representative sample.

Typical samples

Type of organism	Type of trap
slow invertebrate, e.g. snail	hand pick
fast-moving invertebrate, e.g. ground beetles	pitfall trap
small mammals	Longworth trap

- Each animal is marked, using something (e.g. a UV marker pen) that is waterproof, non-toxic and does not affect its differential survival rate, e.g. make it more vunerable to predators.
- The animals are released back into their community.
- A day or two later, a second large sample of animals are caught and the number of marked individuals counted.
- The Lincoln index can then be used to calculate an estimate of the size of population.

$$\textbf{population size} = \frac{n_1 \times n_2}{n_m}$$

n_1 = number caught and marked first time
n_2 = number caught the second time
n_m = number of marked individuals caught second time

You might be asked to explain how you would estimate the population of a small animal you have not studied. Look at the information about the animal in the question – an obvious trapping method should suggest itself.

✓ Quick check 2

Quick check questions

1. Dandelions are weeds that often grow in lawns. Explain how you would estimate how much of a lawn was covered by dandelions.
2. You catch 50 woodlice, mark them and release them again. The next day, you catch 50 woodlice again and find that 15 are marked. Estimate the size of the population of woodlice.

B

HB

Diversity

Diversity depends on the **number of species** (**species richness** of a community) in an ecosystem and the **abundance of each species** – the **number of individuals of each species.**

The populations an ecosystem can support depends on abiotic and biotic factors. Growth of populations depend on **limiting factors**.

Abiotic factors are important because:

- physiological adaptations of organisms only allow them to live in a certain range of e.g. pH, light and temperature – it is part of what defines their niche.
- they affect sizes of populations but are often not greatly changed themselves by sizes of populations (density independent factors).

Abiotic factors often have their effects through enzyme activity, photosynthesis and respiration.

Climatic factors include:

- seasonal changes in temperature, daylength and rainfall,
- longer term changes due to climate changes.

Inorganic ions such as **nitrate** and **phosphate** often limit plant growth.

- Plants are **primary producers**, affecting all populations in a community.

Biotic factors – interactions between organisms

- **Intraspecific competition** occurs between members of the same species
- E.g. for space – a patch of soil to grow on, or a nesting site and food.
- **Interspecific competition** – different species needing the same resource – at the same trophic level.
- Plant species compete for light, herbivore species compete for plants, or carnivore species compete for prey.
- Predation – a predator is a limiting factor on growth on the population of its prey and the prey is a limiting factor on the predator population.

Less extreme environments

- Here the **diversity of organisms is usually high.**
- **Biotic factors dominate and abiotic factors are not extreme.**
- Many species have adaptations that allow them to survive, including **many plants/producers.**
- **Food webs are complex**; with many inter-connected food chains.
- This results in a **stable ecosystem** because, if the population of one species changes, there are alternative food sources for populations of other species.

In less extreme environments organisms often change abiotic factors, making conditions favourable for more organisms to colonise the environment. This is especially true of soil factors.

Extreme environments

- In **extreme environments, diversity is low.**
- **Abiotic factor(s)** are **extreme** and **few species** have adaptations allowing them to survive – **abiotic factors dominate.**

B

HB

- **Food webs are relatively simple**; with few food chains, or connections between them – because **few producers** can survive.
- This can produce an **unstable ecosystem** because, a change in the population of one species can cause big changes in populations of other species.

Quick check 1

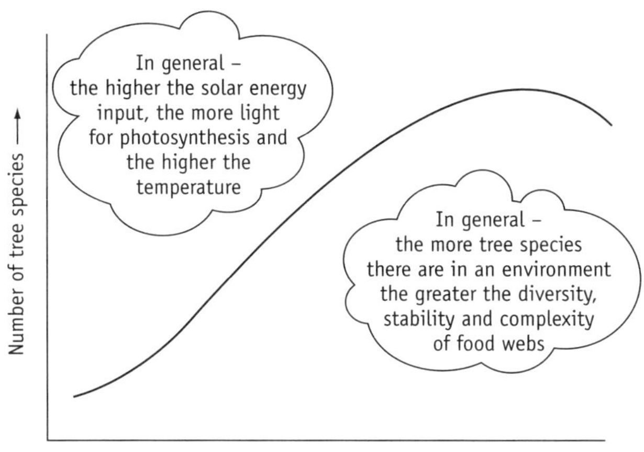

Relationships between light, temperature, diversity and food webs

Quick check 2

Index of diversity

There are many different formulae for calculating a measure of diversity. They usually **take into account the number of species present and the number of individuals of each species**. The formulae are arranged to give **larger numbers for greater diversity.**

- The formula in the specification is derived from the **Simpson's Index** and is given as:

$$d = \frac{N(N-1)}{\sum n(n-1)}$$

d = diversity index
N = total number of organisms of all species
n = total number of organisms of each species

Quick check 3

Quick check questions

1. Oak woodlands are usually described as stable ecosystems. Explain the characteristics you would expect this ecosystem to have.
2. Suggest why the number of tree species increases to a peak and then falls as the annual amount of solar energy increases.
3. One of the captions to the graph states, "In general, the more tree species there are ... the greater the diversity ..." Explain what other information is needed to confirm this statement.

B

HB

Succession and climax communities

Areas of bare land occur where new soil is formed, or where a community has been destroyed. If bare land is undisturbed, a process of **succession** will take place. Communities will form and be replaced by later communities, until a **climax community** is established which remains stable over a long period of time. A **sand dune succession** is described here, as the one example you have to be able to describe. The same principles would apply to any other succession you have studied.

- The first community on bare land consists of **colonisers**, **pioneer species** of **herbaceous (non-woody) plants** and organisms that feed on them.
- These species are able to establish themselves quickly and are adapted to withstand unfavourable abiotic environmental factors.
- In sand dune complexes, new dunes are formed as small mounds of loose sand deposited by the sea. These 'embryo dunes' give poor anchorage for plants, the sand does not hold water and is poor in nutrients.
- Marram grass is a species of herbaceous plant that can colonise these dunes, because it has many adaptations to living in dry conditions (xerophytic adaptations).
- Over time, organisms of the 'marram community' die and are decomposed by bacteria and fungi.
- This adds organic material, humus, to the sand and nitrogen compounds are returned to the soil.
- Humus improves the water-holding capacity of the soil and binds the sand particles together more firmly.
- Plants also **change abiotic factors** such as temperature, wind speed and light by providing shade and acting as wind breaks.
- These changes in the soil **make the environment less hostile/severe**.
- They also produce new niches for new species to enter and form a new community, such as grasses.
- These grow fast and close together, **out-competing** the colonisers to form a 'grass community'.
- The grass community continues to alter the environment, making it possible for a 'shrub community' to become established – dominated by species such as bramble or sea-buckthorn.
- Finally a 'tree community' replaces the shrub community – dominated by species such as birch or oak.
- Changes from one type of community to the next happen because of inter-specific competition. For example, trees finally dominate due to their height and ability to exclude light from shrubs and most herbaceous plants.

The changes and improvements in soil factors are very important and are brought about by plant communities. The availability of water and mineral ions is critical to plant growth and other organisms rely on plants!

Ⓢ Humus holds water and dissolved mineral ions which can be recycled. This makes water more available for plant roots to take up by osmosis along a water potential gradient. More mineral ions are available for uptake by active transport. Higher temperatures give higher rates of enzyme activity and higher rates of respiration – providing more energy for growth, reproduction, etc.

B

HB

The climax community that finally forms depends on abiotic factors – a **succession can stop before it reaches a tree community**. This is usually due to abiotic factors which are too great for a community to alter significantly. An example is seen in the changes in plant communities as you go from a valley to the top of a high mountain.

✓ Quick check 1, 2

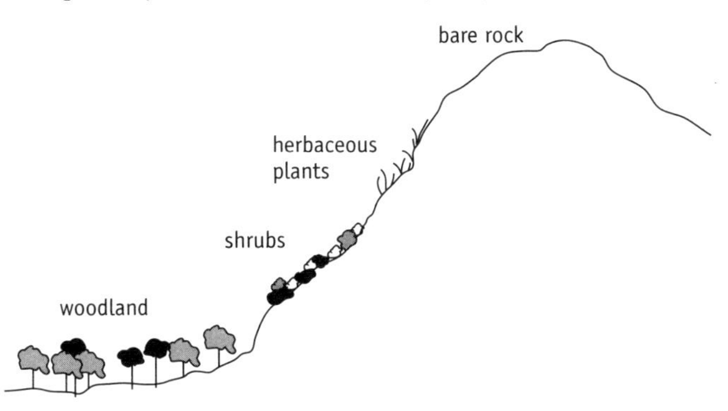

Climax communities

- In the valley, the abiotic factors are not severe and succession goes through to a climax woodland community (like oak).
- Higher up it is too cold, exposed to wind, or the soil is to poor to allow tree seedlings to grow and succession stops at a climax community dominated by shrubs (like heather).
- Higher still, conditions are too extreme for shrubs and the climax community is made up of herbaceous plants (like grasses).
- On the mountain top the conditions are too extreme for plants and bare rock is found.

A question may concern a different succession to the one you studied. You have to apply the ideas you learnt to the new example. Do not ignore the question and write about what you studied!

✓ Quick check 3

? Quick check questions

1. Describe what is meant by a succession.
2. Explain how colonisation of bare soil by herbaceous plants can make it possible for woody plants to grow.
3. Suggest why marram grass does not grow on old sand dunes which have a woodland community growing on them.

B

HB

Photosynthesis

Plants absorb light energy and change it to chemical energy in bonds holding organic molecules together. Photosynthesis takes place in two stages, the light-dependent and the light-independent reactions.

Ⓢ Photosynthesis is the route by which energy enters food chains.

Light-dependent reaction

The light-dependent reaction uses light energy to make some ATP and a reduced coenzyme.

- Light energy is absorbed by electrons in **chlorophyll.**
- Some **excited electrons** gain enough energy to **leave chlorophyll.**
- These electrons are replaced from the **photolysis of water** – water molecules break down, releasing **H^+ ions**, **electrons** and **oxygen.**
- Oxygen is lost as a waste product.
- This **oxygen is valuable**, because it is needed for **aerobic respiration** (see page 32).

Photosynthesis put the 21% oxygen in the atmosphere that we have today and renews it as it is used for respiration by organisms.

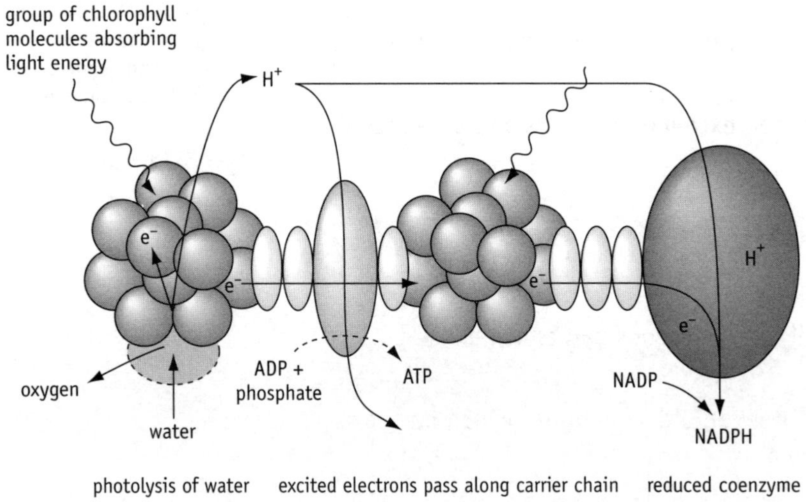

photolysis of water   excited electrons pass along carrier chain   reduced coenzyme

Light-dependent reaction in granum membrane

- The excited electrons **reduce** the first of a chain of **electron carriers – absorbed light energy has been converted into chemical energy (reducing power)**.
- Electrons lose energy as they pass along the carrier chain and this is linked to making ATP.
- At the end of the **carrier chain**, electrons **reduce coenzyme NADP to form NADPH.** (This is where electrons are reunited with the H^+ ions from water.)
- **The reduced coenzyme is a source of reducing power which is used in the light-independent stage of photosynthesis.**

A coenzyme is a non-protein which has to be present for an enzyme to work.
The reduced coenzyme produced here becomes a source of hydrogen for an enzyme that adds hydrogen to a molecule.

Chlorophyll and the proteins of the electron carrier chain are in the membranes of the **grana** of the chloroplast.

✓ Quick check 1, 2

Make sure you know where the two stages of photosynthesis take place in the chloroplast.

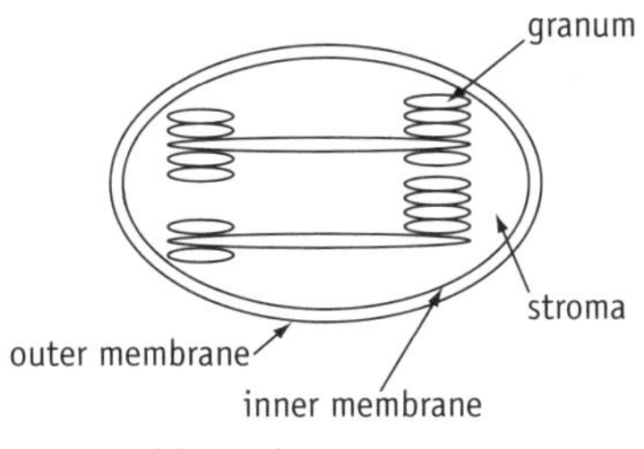

Chloroplast structure

Light-independent reaction

ATP and **NADPH from the light-dependent stage** are used in the **stroma** in the **light-independent** stage. Carbon dioxide (hydrogen carbonate ions) diffuses into the stroma along a concentration gradient.

- Carbon dioxide reacts with an **acceptor** molecule, **ribulose bisphosphate** (RuBP, a five-carbon sugar).
- This produces two molecules of three-carbon **glycerate 3-phosphate** (GP)**.**
- This is **reduced** to **triose phosphate (a sugar)**.
- This is converted to **sugar (glucose)** in a **reduction reaction.**
- The hydrogen for this comes from **NADPH.**
- To lower the activation energy for the reactions, **ATP** is used to phosphorylate the reactants**.**
- Some of the sugars produced are used by the plant, the rest are used to **regenerate ribulose bisphosphate** in **the Calvin cycle.**

Ⓢ Glycerate 3-phosphate is an acid that is reduced to a sugar by NADPH

Ⓢ This stage relies on enzymes and its rate is affected by factors that affect enzymes, e.g. temperature.

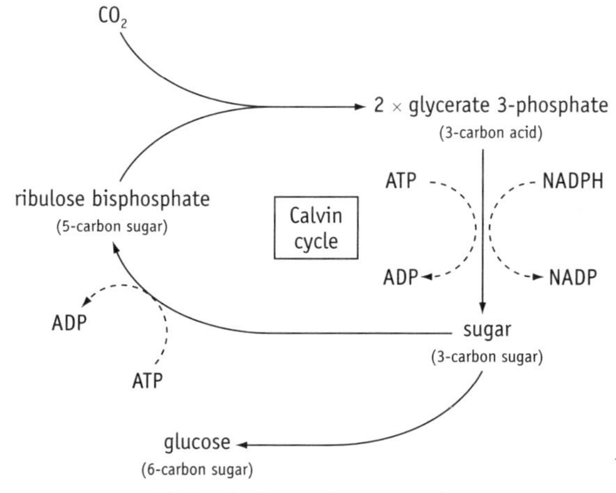

Light–independent reaction

Don't try to learn any more details of the biochemistry of photosynthesis. What you need to know is here!

✓ Quick check 3, 4

? Quick check questions

1. Explain why water is needed for photosynthesis.
2. Explain how light energy is used in photosynthesis.
3. Radioactive carbon dioxide was given to a plant. Suggest the order in which molecules in the chloroplasts would become radioactive.
4. Explain why the light-independent stage of photosynthesis stops very quickly if a plant is put into the dark.

B

HB

Energy transfer

Photosynthesis is the means by which energy enters an ecosystem. Plants are **autotrophs**, making biological molecules from sugars made in photosynthesis and inorganic mineral ions.

- **Primary producers – plants** – provide food for all other organisms.
- **Food chain** – a series of feeding relationships – starts with a primary producer.

tertiary consumer (carnivore) *fourth trophic level*

⇑

secondary consumer (carnivore) *third trophic level*

⇑

primary consumer (herbivore) *second trophic level*

⇑

primary producer plants *first trophic level*

Food chain

- **Feeding transfers energy** – arrows point in the direction of energy flow.
- Each level in the food chain is a **trophic level** (energy level).
- Little energy is transferred from one trophic level to the next, because energy is **dissipated (lost)** by organisms at each trophic level through:
  - **respiration**, which provides energy used by organisms for, e.g. movement, growth, reproduction,
  - **heat** energy from respiration,
  - **excretion** of waste products like carbon dioxide and urea,
  - **decomposition** of dead organisms which were not eaten.
- **Decomposers** – bacteria and fungi feed on dead organisms.
- Eventually, all energy is **dissipated back into the environment** as heat energy.
- **Efficiency of energy transfer between trophic levels** can be calculated.

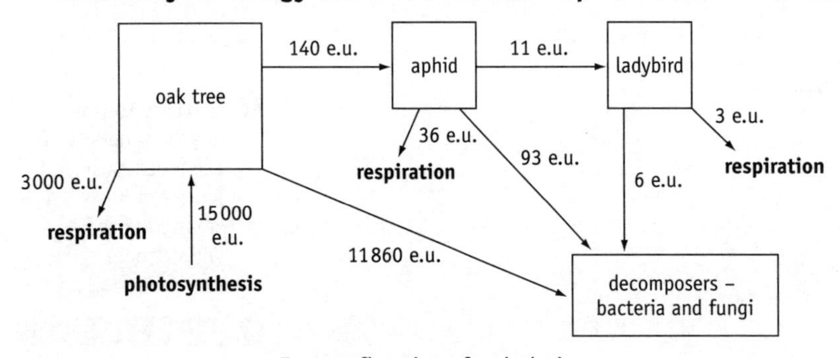

Energy flow in a food chain

Example: What percentage of the energy input from photosynthesis goes to decomposers?

Input from photosynthesis = 15 000 e.u.

Energy input into decomposers = 11 860 + 93 + 6 = 11 959 e.u.

$$\% \text{ of energy going into decomposers} = \frac{11\,959 \times 100}{15\,000} = 79.7\%$$

Ⓢ In photosynthesis light energy is converted to energy in chemical bonds holding sugar molecules together. All organisms carry out respiration all of the time to produce ATP. Heat energy is lost during the process.

Some organisms occupy more than one trophic level. Humans are omnivores – they eat plants and animals.

✓ *Quick check 1, 2*

- Food chains are part of **food webs** - **interconnected food chains**.

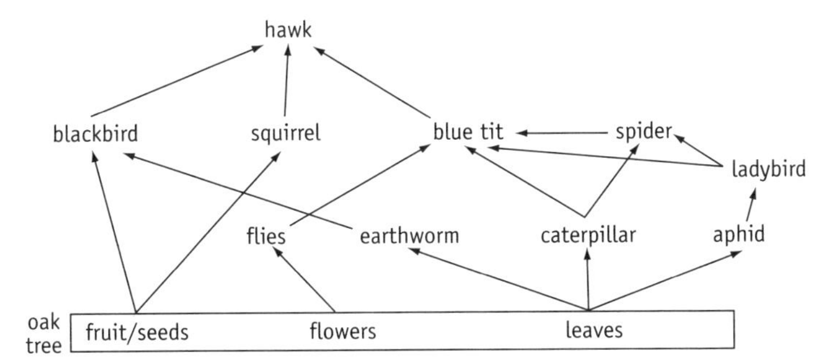

Simplified food web

Ecological Pyramids

Pyramid of number – **number of organisms** at each level of a food chain/web **at a specific time**. Primary producers are the base of the pyramid. The size of a block represents the number of individuals at each level.

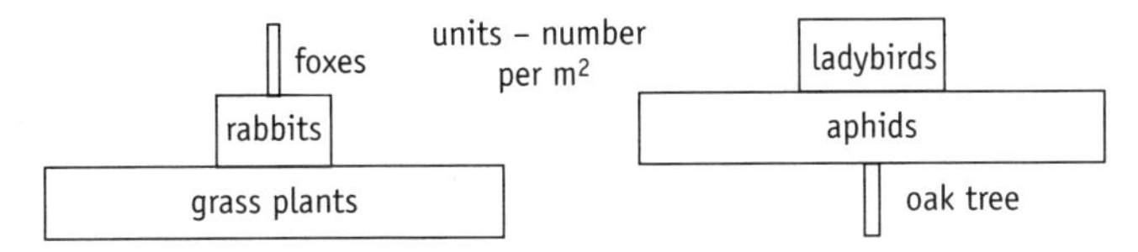

Pyramid of number

- This often gives a misleading impression of the energy transfers. For example, one oak tree will support very large numbers of consumers.

Pyramid of biomass – the **mass of organisms** in each level of a food chain at a **specific time, in a given area** (e.g. **Kg m^{-2}**).

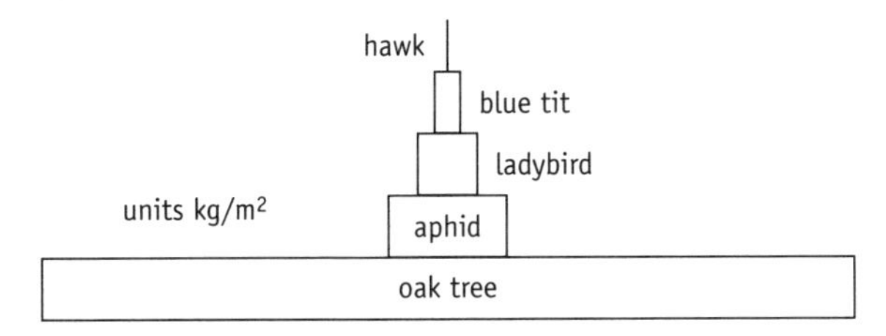

Pyramid of biomass

It is fairly easy to count organisms in an area. It is harder to find their mass – e.g. of an oak tree – without destroying them! Finding energy content is destructive and involves fairly complex lab work. It still doesn't give energy flow!

- This gives some idea of energy present but it does give the **rate** at which matter/energy passes between trophic levels.

Pyramid of energy shows **energy present** at each trophic level, for a **particular area**, for a **particular period of time** (e.g. **kJ m^{-2} yr^{-1}**).

Quick check 3, 4

Quick check questions

1. Suggest why energy losses mean that food chains can only involve five or six trophic levels at most.
2. What percentage of the energy going from producers to ladybirds goes to decomposers?
3. Explain the differences in units used for pyramids of number, biomass and energy.
4. Suggest why it is difficult to get data for pyramids of energy.

B

HB

Energy supply

Chemical energy

All cells (and organisms) respire all the time, including plant cells. In **respiration** the chemical (covalent) bonds between the atoms of respiratory substrates (organic molecules) are broken. This releases energy that can be used to make **ATP**. ATP contains **chemical energy** that can be used for biological processes.

- **Photosynthesis** in plants makes organic glucose molecules from inorganic carbon dioxide and water molecules.
- These **synthetic reactions**, make larger, more complex molecules from smaller, simpler molecules.
- **Light energy** is absorbed by chlorophyll and provides the energy input needed for the synthetic reactions.
- **Respiration** in plant cells uses a lot of the glucose produced by photosynthesis.
- Animals get respiratory substrates by **feeding** and plants make them in **photosynthesis**.

Ⓢ Plants are producers in food chains. They can make all the different types of organic molecules found in cells from sugars made in photosynthesis. Other organisms get these molecules from plants by feeding.

✓ *Quick check 1*

ATP

An ATP molecule is made by adding an inorganic **phosphate group** to **ADP**.

- This reaction needs an input of energy and is linked either to energy released when chemical bonds in glucose (or other carbohydrates, lipids or amino acids) are broken in respiration, or to the light-dependent reaction of photosynthesis.
- ATP can transfer a phosphate to another molecule – **phosphorylation**.
- The phosphorylated molecule is more reactive and this **lowers the activation energy** needed for a reaction involving an enzyme.
- **ATP becomes ADP again when it loses its phosphate group**.

Ⓢ Enzymes also lower the activation energy of reactions.

✓ *Quick check 2*

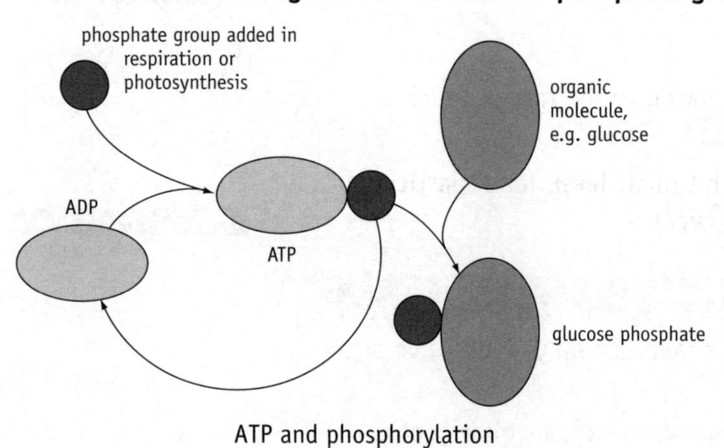

ATP and phosphorylation

A lot of the biochemistry you have already learnt is simplified. Large molecules are actually made from smaller, phosphorylated molecules. For example, starch is made from glucose phosphate and proteins are made from phosphorylated compounds of amino acids.

Oxidation and reduction

- Many reactions in respiration and photosynthesis involve oxidation and reduction. There are **three types of oxidation reaction**:
  - **loss of electrons,**
  - **removal of hydrogen**,
  - **combination with oxygen**.
- Oxidation is linked to **reduction**; oxygen is lost or hydrogen or electrons are gained by an atom, molecule or ion.
- The reduced substance is a source of chemical energy (**reducing agent**).

Make sure that you remember that the removal of hydrogen from a molecule is an oxidation!

Respiratory substrates and respiratory quotient

A **respiratory substrate** is a molecule which is oxidised during respiration. Glucose is one respiratory substrate but there are others.

- Resting muscle uses **fatty acids** from **triglycerides/lipids** as the main respiratory substrate.
- During exercise, muscle switches mainly to **glucose.**
- Muscles contain a lot of **glycogen**, which can be converted to glucose.
- Surplus **amino acids** from **protein** are broken down in the liver to give **organic acids** which can be respiratory substrates.

The **respiratory quotient** is:

$$RQ = \frac{\textbf{carbon dioxide produced}}{\textbf{oxygen used}}$$

The **theoretical RQ values** for different respiratory substrates are:

- **carbohydrate** = **1** – one O_2 used in **aerobic respiration** for each CO_2 produced,
- **lipid = 0.7** – lipids contain little oxygen/more hydrogen compared to carbohydrates, so more O_2 needed,
- **protein = 0.9**.
- Animals usually use more than one type of respiratory substrate, so their RQ will be between the values above.
- **Humans** rarely depend on protein, so their RQ is usually about 0.85 – reflecting use of carbohydrate and lipid.

Anaerobic respiration produces a lot of CO_2 but doesn't use oxygen – giving a number greater than one! A plant cell may use all or most of the CO_2 produced in respiration in photosynthesis and release O_2!

Quick check 3

Quick check questions

1 Students often write that animals respire but plants photosynthesise. Explain what is wrong with that statement.

2 Explain why ATP is necessary for reactions in cells to take place.

3 Suggest a likely value for the RQ of a domestic cat.

B

HB

Respiration

The breaking of chemical (covalent) bonds in glucose (other carbohydrates, lipids or amino acids) releases energy. Some of this energy is used to make ATP and the rest appears as heat. Glucose is broken down in a series of reactions involving **oxidations,** linked to **reduction** of **coenzymes.** Reduced coenzymes are the source of the energy (**reducing power)** used to make ATP.

Ⓢ Glycolysis does not need oxygen. It is always the first stage of respiration. In anaerobic respiration it is the only stage.

Glycolysis

This is the first stage of respiration and takes place in the **cytoplasm. Six-carbon** glucose is oxidised into **two, three-carbon** molecules of **pyruvate.**

- Glucose is phosphorylated with two phosphate groups from two ATP molecules.
- Glucose is **oxidised** by **removing hydrogen** and **ATP is produced.**
- Hydrogen is accepted by a **coenzyme, NAD**, forming **reduced NAD.**
- This oxidation of glucose produces two molecules of **pyruvate.**
- There is a **net gain of two ATP** (four are made – two are used at the start) molecules.

✓ Quick check 1

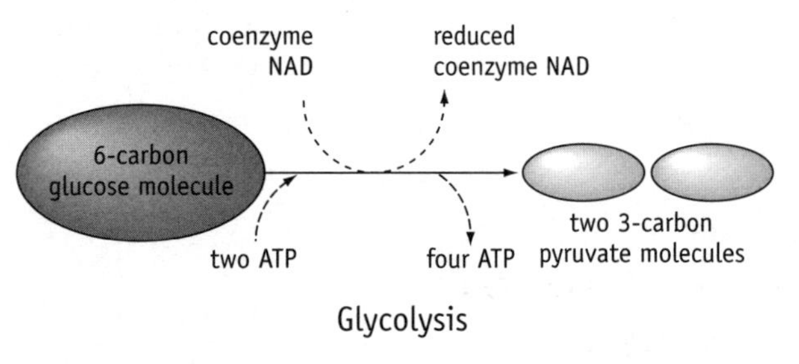

Glycolysis

Krebs cycle

In aerobic respiration, pyruvate from glycolysis enters the Krebs cycle in the matrix of the **mitochondrion.**

This **produces reduced coenzymes NAD and FAD.**

Don't try to learn any more details of the biochemistry of respiration. What you need to know is here!

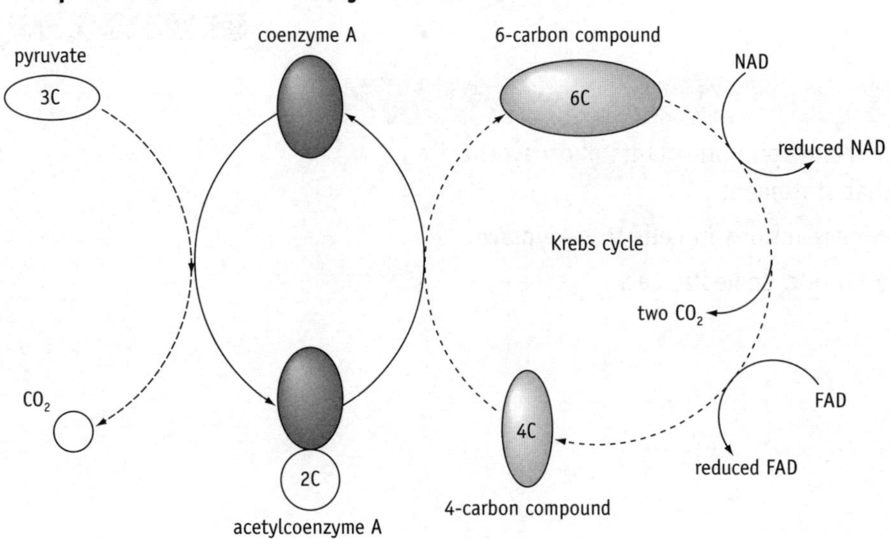

Krebs cycle

- **Pyruvate** reacts with **coenzyme A,** producing **acetylcoenzyme A** (carrying **two carbon atoms** from pyruvate) and releasing carbon dioxide.
- Acetylcoenzyme A reacts with a four-carbon compound, forming a **six-carbon molecule.**
- **This is oxidised by removing hydrogen**.
- Hydrogen is accepted by coenzymes, to give **reduced coenzymes NAD and FAD** ($NADH_2$ and $FADH_2$) – the **main products of the Krebs cycle.**
- Carbon dioxide is released as a waste product and the four-carbon molecule that reacts with acetylcoenzyme A is reformed.
- A small amount of ATP is made by the Krebs cycle.

Make sure that you answer the question set! **Don't** be tempted to write everything you have memorised about respiration – especially details of biochemistry not in the specification!

Quick check 2

Oxidative phosphorylation

This is associated with the inner mitochondrial membrane (cristae) and produces almost all the ATP in aerobic respiration.

- Reduced NAD and FAD from glycolysis and the Krebs cycle reduce the first protein in an **electron transport chain**.
- The hydrogen from the coenzymes gives **H^+** and an **electron.**
- The **electron loses some energy** as it passes down the transport chain.
- Some of this energy is used to **phosphorylate ADP to ATP**.
- This is the **oxidative phosphorylation** of ATP.
- Electrons coming off the end of the electron transport chain are accepted by **oxygen** and joined by H^+ to make water (another waste product of respiration).

Ⓢ The heat produced in respiration is energy which is lost to the environment from organisms and food chains.

Quick check 3, 4

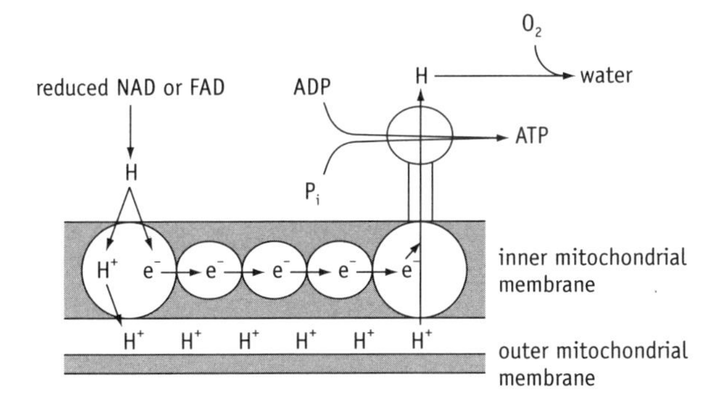

Oxidative phosphorylation

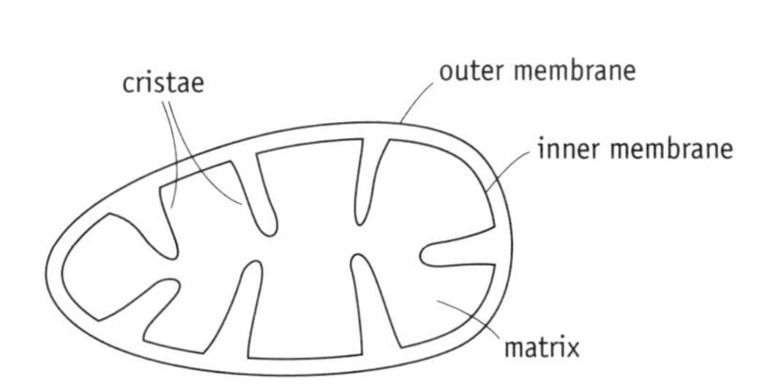

Structure of a mitochondrion

Quick check questions

1. Describe the oxidation of glucose in glycolysis.
2. Explain how the Krebs cycle is linked to ATP production.
3. Explain how ATP is produced by oxidative phosphorylation.
4. Cyanide is a poison that stops the reactions combining oxygen, electrons and H^+ to make water. Suggest how cyanide kills.

B

HB

Nutrient cycles

In an **ecosystem** interactions occur between **biotic** (living) and **abiotic** (non-living) things. Chemical elements constantly enter and leave the biotic environment.

- **Carbon** enters the biotic environment when **carbon dioxide** (inorganic) is used to make **sugars** during **photosynthesis**.
- It returns to the abiotic environment as carbon dioxide; during **respiration**.
- **Microorganisms decompose** dead remains of organisms, returning **inorganic ions** to the environment for other organisms to re-use.
- **Bacteria and fungi** are **microorganisms** and important **decomposers**; obtaining their food by **saprotrophic nutrition**.
- They secrete digestive enzymes onto their food to digest large, insoluble food molecules into smaller, soluble molecules that they can absorb through their cell surface membranes.
- Many inorganic ions are in short supply in the abiotic environment and are limiting factors in the growth of producers.

Ⓢ You need to know how plant roots take up water and mineral ions.

Ⓢ Decomposition happens faster at warmer temperatures, because the enzymes of bacteria and fungi work faster – up to their optimum temperature.

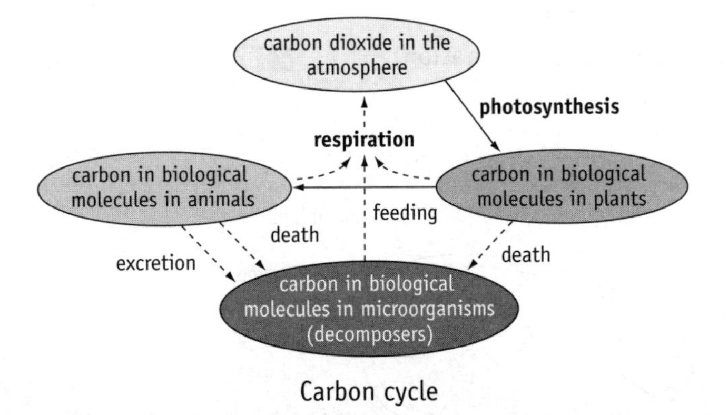

Carbon cycle

- **Microorganisms** carry out **respiration,** releasing carbon dioxide to the atmosphere.

✓ Quick check 1

The nitrogen cycle

- Plants need the element nitrogen for **amino acid and protein synthesis** (and synthesis of nucleotides).
- Plants can not use molecular nitrogen gas from the atmosphere.
- **Nitrogen fixation** – atmospheric nitrogen is converted into forms plants can use (ammonia, and **nitrate ions (NO_3^-)**), by **nitrogen-fixing bacteria.**
- ***Rhizobium*** is a bacterium in root nodules of leguminous plants (e.g. clover), that fixes nitrogen into ammonia, which the plant can use.
- **Nitrifying bacteria** convert ammonia to nitrites and nitrates – **nitrification.**
- **Denitrifying bacteria** break down nitrates, releasing nitrogen gas – **denitrification.**
- Consumers in food chains depend on plants for their amino acids.

- Animals covert surplus amino acids to ammonia and organic acids – **deamination**.
- Ammonia is excreted, or urea, or uric acid (nitrogenous wastes).
- Ammonia is produced when dead organisms are **decomposed.**

✓ Quick check 2, 3

Do not try to learn the proper names of the various types of bacteria in the nitrogen cycle.

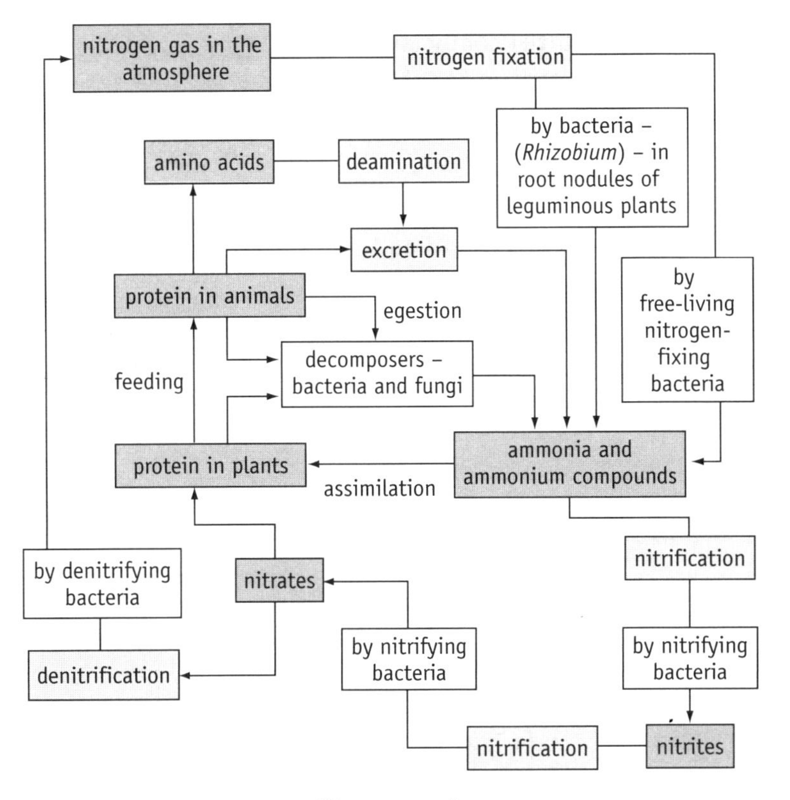

Nitrogen cycle

? Quick check questions

1 Explain the significance of photosynthesis and respiration in the carbon cycle.

2 Explain the importance of microorganisms in the recycling of materials in ecosystems.

3 **(a)** Explain how the nitrogen gas from atmosphere gets into proteins in plants.

(b) Suggest why nitrates in the soil are often the limiting factor in the productivity of an ecosystem.

B

HB

Deforestation

Human activities often affect whole ecosystems. There are potential conflicts of interest between the need/wish to produce things useful to humans in the short term and the conservation of ecosystems in the long term.

- **Forests** are the natural **climax communities** in many parts of the world.
- They have **high diversity**, with **complex food webs.**
- Humans have been clearing areas of forest for thousands of years – leading to **deforestation** over large areas of Europe, Asia and North America.
- Recent and present deforestation affects mainly **tropical rain forests.**
- Deforestation in Britain has been taking place for thousands of years and has only been slightly reversed in the last few decades.

Tropical forests grow where abiotic factors are not extreme and biotic factors dominate. This leads to complex food webs and a stable ecosystem – unless humans interfere!

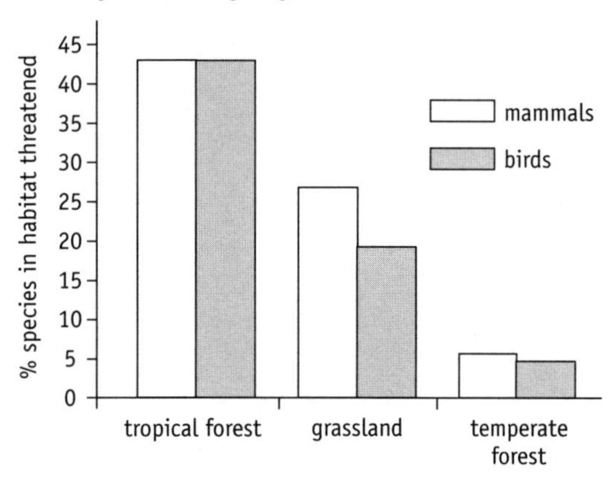

Percentage of species threatened in tropical forests, compared to temperate forests and grasslands

Agriculture

- Growth in human populations is increasing demand for land for farming.
- High demand for animal protein (e.g. beef) in rich countries has increased deforestation in poorer countries, to provide grazing land.

Diversity

- Deforestation causes local extinction of species of trees.
- This particularly affects hardwoods which are in demand for timber in rich countries and softwoods used for making paper.

Loss of trees:

- removes the bases of many food webs,
- removes the habitats of many other species,
- causes local extinction of other populations, or reduction in their size,
- reduces the number of species present and numbers of individuals present,
- reduces diversity,
- leads to a lower biomass and productivity per hectare.

Reducing diversity produces a less stable and more extreme environment, where abiotic factors also become more extreme.

Quick check 1

Carbon and nitrogen cycles

- Loss of trees (and other plants in a community) means **less photosynthesis**.
- Deforestation usually involves **burning** unwanted trees, and expanding human populations burn more wood for fuel.
- Less carbon dioxide is removed from the atmosphere and more is added.
- This adds to the problem of **global warming**.
- In forests, most of the **nitrate ions** (and other mineral ions) absorbed by plants come from decomposition of organic remains – the ions are **recycled**.
- Many of the decomposing fungi live in association with the roots of trees.
- The soil itself is often a poor source of mineral ions.

In the warm, moist environment of tropical forests the action of decomposers is very rapid – leading to rapid recycling of nutrients. The warmth speeds up enzyme activity.

Deforestation results in:

- reduced input to the nitrogen cycle,
- slower and less recycling of nitrates (and other ions),
- increased loss of nitrates by leaching.

The soil loses fertility and can support lower numbers and fewer species of plants – lower diversity.

✓ Quick check 2

Conservation of forests

Many attempts are being made to encourage **sustainable use of forests.**

This involves measuring and comparing yields and profits from deforestation with alternative uses.

Example: Tropical forests of the Amazon basin.

The figures given applied at the time the study was done – they will be different now.

- Felling one hectare gave a one-off crop of wood worth about £670.
- Using cleared land for cattle produced about £100 per hectare per year.
- The land often loses its fertility after a few years.

If the forest is kept it can yield:

- £280 per hectare per year in rubber and fruit – indefinitely,
- income from tourism,
- income from medicinal plants – many yet to be discovered,
- gene pools of wild relatives of domesticated organisms – which may be used as a source of genes/alleles in selective breeding or genetic engineering.

✓ Quick check 3

? *Quick check questions*

1. Explain how deforestation leads to lower diversity.
2. Suggest why land cleared of trees usually becomes less fertile.
3. Calculate the profits made over ten years when harvesting rubber and fruit from the forest and compare this with the profit from clearing forest for cattle ranching.

B

HB

Module 5: end-of-module questions

1 The number of chromosomes in a skin cell of a mammal was found to be 44, consisting of 22 homologous pairs.

a What is the : (**i**) haploid number; (**ii**) diploid number in this animal? [2]

b Explain what is meant by a homologous pair of chromosomes. [2]

c Describe how genetic material is exchanged between homologous chromosomes. [2]

d Explain how a constant chromosome number is maintained from one generation to the next in sexual reproduction in a mammal. [3]

2 The table shows the phenotypes and possible genotypes of ABO blood groups.

Blood group phenotype	**Blood group genotype**
A	I^AI^A, I^AI^O
B	I^BI^B, I^BI^O
AB	I^AI^B
O	I^OI^O

a I^A, I^B and I^O are all alleles of the same gene.

i What is an allele? [1]

ii Explain which of these alleles is recessive. [1]

iii Explain which alleles are codominant. [1]

iv Explain the type of variation shown by ABO blood groups. [3]

b A mother with type B blood and a father with type A blood had a baby with type O blood. Use a genetic diagram to explain how the child inherited its blood group. [4]

3 Warfarin is a pesticide used to kill rats. In a population of rats, tests showed that 1% of the rats were resistant to warfarin, even though the population had not been exposed to the pesticide before. These rats were homozygous for a recessive allele of a gene which gave them resistance to warfarin.

a Explain how the resistant allele could be in the population, before warfarin was used. [2]

b 1% of the population was homozygous resistant. Use the Hardy-Weinberg equation to calculate the percentage of rats in the population that were heterozygous for the allele. [3]

c The graph shows what happened to the population when warfarin was used.

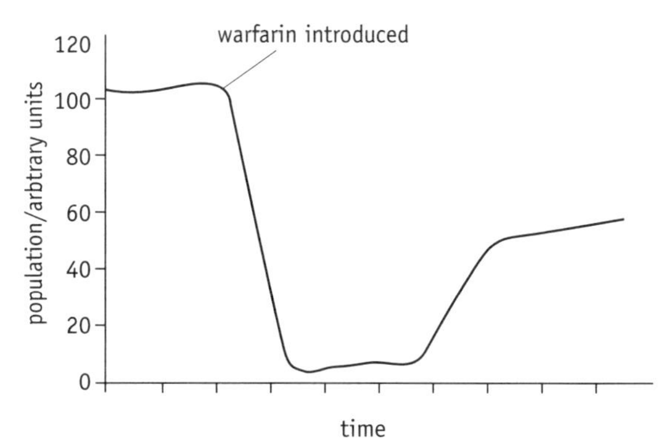

Explain the recovery in the rat population. [4]

4 A woodland ecosystem will contain populations of differing species, which form one or more communities. Within a woodland each species will occupy an ecological niche and the size of its population will vary as a result of biotic and abiotic factors.

In a woodland a forest fire may produce an area of bare soil. After a period of time, this area will usually become woodland again.

a What is meant by: (**i**) a population; (**ii**) an ecological niche? [2]

b Describe the techniques that could be used to compare the vegetation in two areas of woodland.

c A study was made of the diversity in a woodland.

i What information would be needed to calculate an index of diversity for the woodland?

ii Suggest how the diversity in the woodland would compare with a desert community.

5 a **i** Give two products of the light-dependent reaction. [2]

ii Where does the light-dependent reaction take place? [1]

b The diagram shows part of the light-independent reaction.

ribulose bisphosphate $\xrightarrow{CO_2}$ 2 x glycerate 3-phosphate

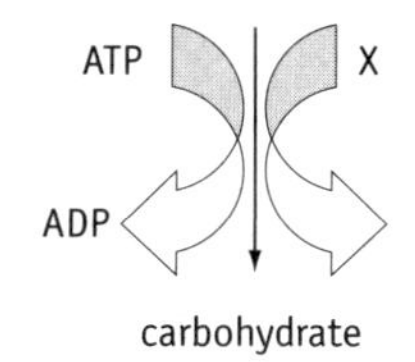

How many carbon atoms are present in a molecule of:

i ribulose bisphosphate,

ii glycerate 3-phosphate? [2]

c Name molecule **X** and explain its role in the light-independent reaction. [2]

B

Module 6: Physiology and the Environment

This module is broken down into 10 topics: Water and flowering plants, Homeostasis*, Functions of the liver and kidney, Gaseous exchange surfaces, Transport of respiratory gases, Digestion and absorption*, Neurones, action potentials and synapses*, Receptors*, Control of behaviour and Metamorphosis and Insect diet.

Some sections* are common to Module 7 (Human Biology) and are marked  HB

Water and flowering plants

- Water enters by the apoplast and symplast pathways.
- It moves through xylem by capillarity, root pressure or cohesion-tension.
- Plants lose water by transpiration – xerophytes' adaptations reduce losses.

Homeostasis  

- Homeostasis gives a constant, optimal, internal environment.
- Temperature and blood glucose are controlled; by negative feedback.

Functions of the liver and kidney

- Different groups of animals excrete different types of nitrogenous waste.
- The liver deaminates amino acids and produces urea.
- The kidney filters urea from blood and excretes it in urine.
- The kidney plays a major part in controlling the water balance of the body.

Gaseous exchange surfaces

- All gaseous exchange surfaces share the same basic characteristics.
- Water is lost from gaseous exchange surfaces but organisms have adaptations to limit losses.

Transport of respiratory gases  

- Haemoglobin carries oxygen in the blood and carbon dioxide is carried mainly as hydrogencarbonate ions.
- Different organisms have different types of haemoglobin with different oxyhaemoglobin dissociation curves.

Digestion and absorption  HB

- Hydrolytic enzymes in the gut digest large biological molecules into small, soluble molecules that can be absorbed.
- The structure of the gut wall is adapted for digestion and absorption.
- Digestive secretions are controlled by the nervous and hormonal systems.

Neurones, action potentials and synapses B HB

- Neurones generate action potentials or nerve impulses.
- Information is carried by the frequency of action potentials.
- Neurones communicate at synapses, where information is also processed.

Receptors B HB

- Receptors convert stimuli into nerve impulses in neurones.
- Pacinian corpuscles respond to pressure and rods and cones in the eye respond to light.

Control of behaviour B HB

- Spinal reflexes give rapid, non-conscious responses to harmful stimuli.
- The autonomic nervous system controls non-conscious responses of many important organs (and systems), such as the heart.
- Taxes and kineses maintain organisms in a favourable environment.

Metamorphosis and insect diet

Some insects show complete metamorphosis; their larval and adult forms have completely different body forms and dietary requirements.

B

Transport in plant roots

The uptake of water and mineral ions by plants from the soil occurs in the roots. These substances are transported across the root tissues and up the plant to the leaves.

Root structure

The piliferous layer has **root hairs** which are thin, permeable, tubular extensions of single **epidermal cells**.

- The exodermis protects against desiccation and the entry of pathogens.
- The cortex consists of parenchymal cells in which starch may be stored.
- The **endodermis** is one cell thick and has impermeable **Casparian strips** made of suberin in its cell walls.

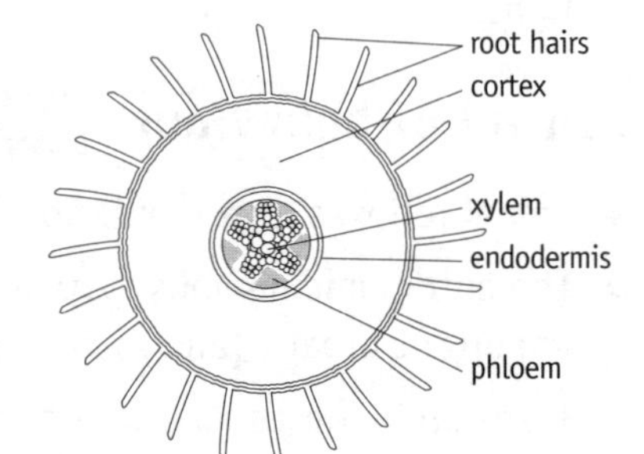

Structure of a root

Phloem

The **phloem transports organic molecules** to or from the root and consists of living cells including sieve elements and companion cells.

- **Sieve elements** possess perforated end walls or **sieve plates**.
- Sieve elements are joined end to end to form **sieve tubes**.
- Mature sieve elements have no nucleus and the cytoplasm contains few organelles.
- Each sieve element has a **companion cell** with a nucleus, dense cytoplasm and many mitochondria.

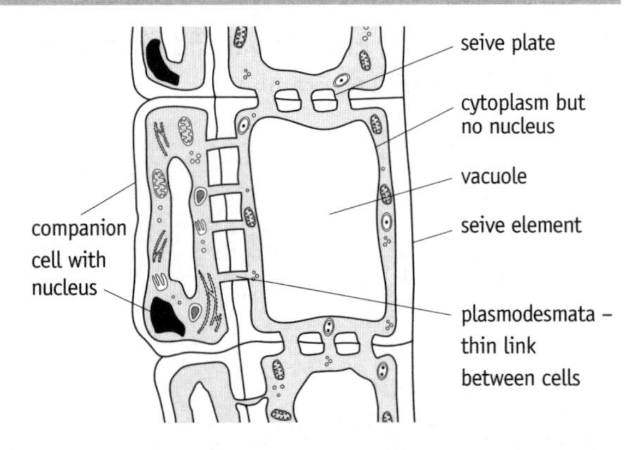

Structure of phloem tissue

In phloem and xylem there are cells joined end to end to make long 'tubes' through which mass flows of liquid and dissolved substances can take place.

Xylem

The **xylem transports water and mineral ions** from the roots to the leaves. It is a non-living tissue consisting of vessels.

- **Vessels** are formed by many xylem cells joining together – their end walls break down to form long, hollow 'tubes'.
- The vessel walls are thickened – so cells do not collapse when water is drawn through them.
- The walls contain impermeable lignin – to keep water in.
- Pits or gaps in the lignin enable water to pass between adjacent vessels.

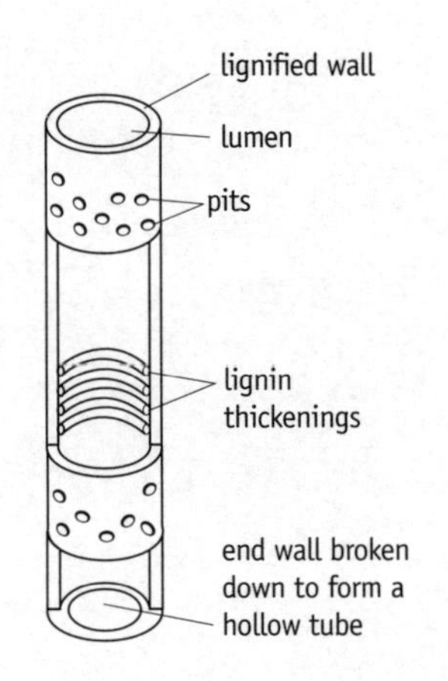

A xylem vessel

Make sure you can label the parts of the primary root.

B

Uptake and transport in roots

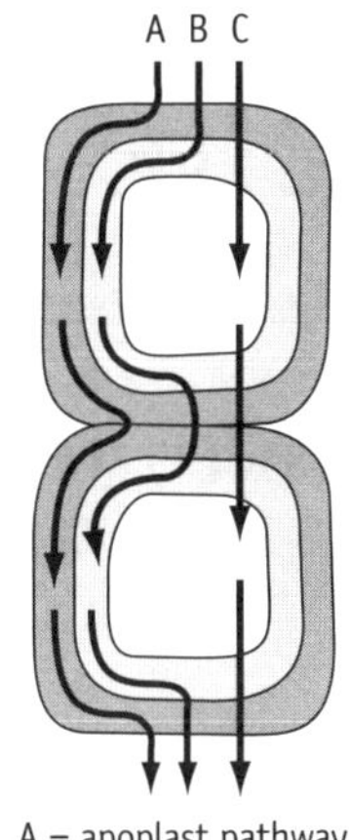

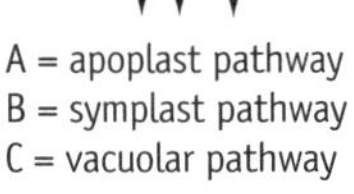

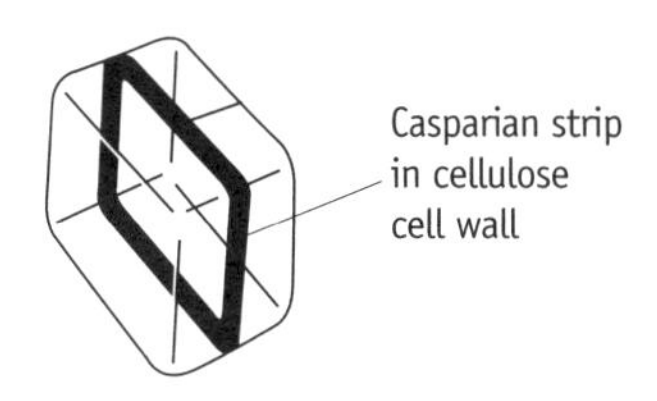

Casparian strip in an endodermal cell

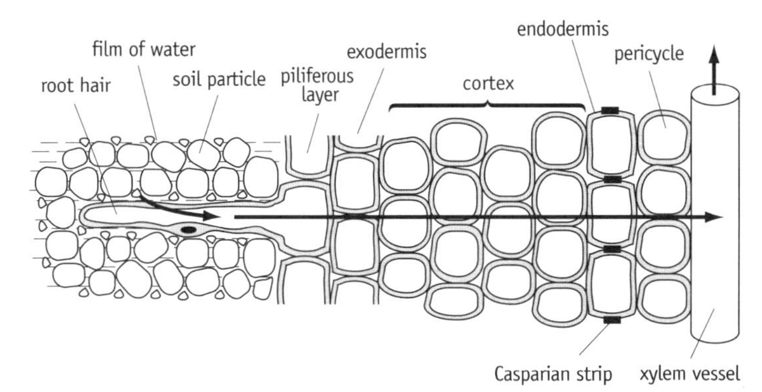

Pathways of transport of water and ions across a root.

- Uptake of water and ions is mainly by the root hairs which provide a large surface area for the absorption of water and ions.
- Ions are absorbed into the root by **diffusion** and **active transport.**
- Water uptake is by **osmosis** along a **water potential gradient** – soil water has a higher water potential (less negative) than root hairs (more negative).
- Water moves along a water potential gradient from root hair cells, across the cortex to the xylem vessels.
- The water potential gradient is maintained by water continually moving up the xylem and by dissolved ions in the xylem sap.

The movement of water across the root cells can occur via three pathways, the apoplast, symplast and vacuolar.

Apoplast pathway

- This is movement of water (and dissolved mineral ions) through the cellulose **cell walls** of adjacent cells and the small intercellular spaces between them.
- Cell walls are fully permeable across the root, except for the **endodermis.**
- Here, the impermeable **Casparian strip** prevents passage of water and dissolved mineral ions through the apoplast.
- They have to cross cell membranes of endodermal cells into the symplast, allowing control of the movement of water and ions into the xylem.
- Endodermal cells actively pump ions into the xylem.

Symplast pathway

- This is movement of water, by osmosis, through the inter-connecting cytoplasm of adjacent cells.
- The water travels through plasmodesmata – thin strands of cytoplasm linking the cytoplasm of adjacent cells.

Vacuolar pathway

- This is the movement of the water through cell vacuoles of adjacent cells.
- The water must cross the cell surface and vacuole membrane.
- The water moves by osmosis along a water potential gradient.

Ⓢ You should know the organelles of a typical plant cell and how their structure is related to their functions.

Quick check 2

The same three pathways account for the movement of water in the leaf from the xylem tissue, across the leaf mesophyll cells and out through the stomata

Ⓢ Cellulose molecules are held together by hydrogen bonds between the long, straight molecules – to form fibrils. There are water-filled spaces between the fibrils which make the cell wall freely permeable to substances in solution.

Quick check 3, 4

Quick check questions

1. Name the parts of the root labelled A, B and C on the diagram.
2. Explain how a low water potential is maintained in the xylem tissue of a root.
3. Describe how water movement occurs by the symplast pathway in a root.
4. How do endodermal cells control water movement into the xylem?

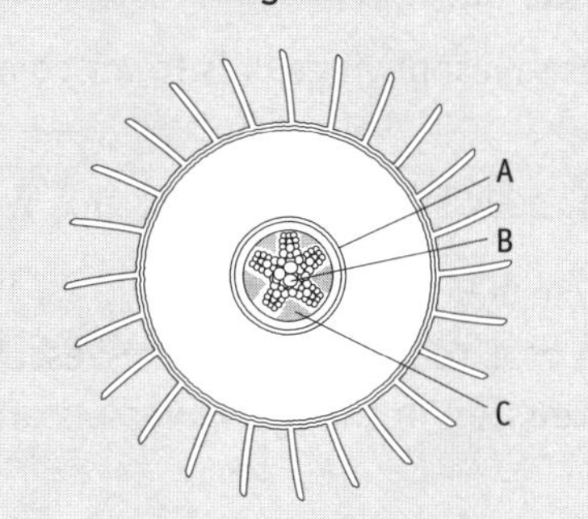

B

Transport in the xylem

In plants, water and mineral ions are transported in the **xylem** and organic substances in the **phloem**.

Ⓢ Always use water potentials, rather than osmotic, solute or other terms. Water diffuses from a solution with a higher (less negative) to one with a lower (more negative) water potential.

The transpiration stream

Water and dissolved mineral ions move from root hairs to stomata through the xylem, as part of the transpiration stream. Theories for the movement of water in the xylem include, **capillarity, root pressure and cohesion-tension.**

There are **cohesive forces** between water molecules; they are attracted to each other by weak **hydrogen bonds** between hydrogen and oxygen atoms. They are also attracted to other 'polar' molecules; e.g. sugars and proteins.

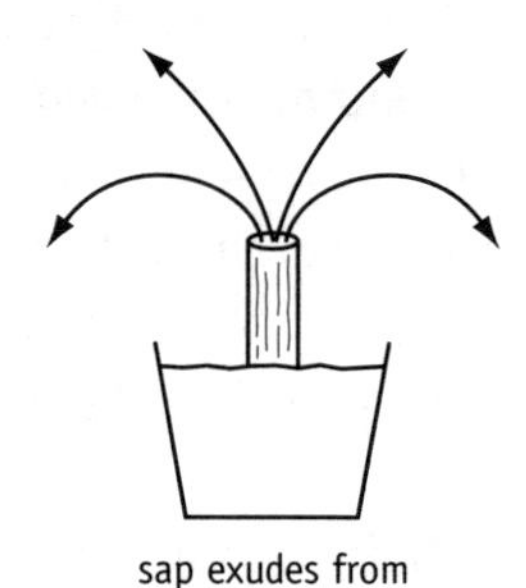

Capillarity in a stem

Capillarity

- Water molecules are attracted to sugar molecules making up the cellulose cell walls of the very narrow xylem vessels.
- Surface tension forces due to cohesive forces between water molecules then make water rise up the xylem – by capillary attraction.

Root pressure

- This is **positive hydrostatic pressure**, seen if sap flows from a freshly cut root stump.
- **Active transport of mineral ions** by endodermal cells **lowers the water potential** of the xylem.
- Respiratory inhibitors, low temperatures and lack of oxygen inhibit this process – by stopping active transport.
- Water moves into the xylem by **osmosis** – raising the hydrostatic pressure.
- Root pressure is not great enough to transport water to the top of trees.

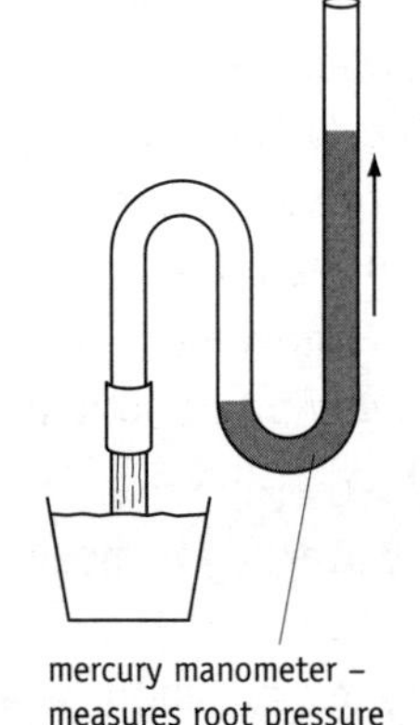

Root pressure in a cut stump of a plant.

Root pressure is not found in all plants and is considered to be of secondary importance compared to cohesion tension.

✓ Quick check 1

Cohesion-tension hypothesis

- Solar heat energy causes **transpiration**/evaporation of water from leaves.
- **Water evaporates** from mesophyll cells next to air spaces and diffuses out through the stomata into the air.
- The water potential of these mesophyll cells is lower compared to inner mesophyll cells.
- Water moves from adjacent cells by osmosis along this **water potential gradient**.
- Movement of water is by the apoplast, symplast and vacuolar pathways.
- This water potential gradient extends to the xylem vessels, drawing water from the xylem and creating a **tension** in the xylem vessels and 'pulling up' water and dissolved ions.

Ⓢ Active transport depends on protein carriers which are part of the fluid mosaic structure of the cell membrane. Their tertiary structure allows them to bind to a specific ion.

B

- Water in the xylem forms a **continuous column** from the leaves to the roots. If the top of the column is 'pulled up', the whole column moves up.
- A continuous water column is maintained by **cohesive** forces and **adhesive** forces – cohesive due to hydrogen bonding between water molecules – adhesive due to attraction of the water molecules to the xylem walls.
- Water moving up from xylem in the roots maintains a water potential gradient across root cortex cells, for water uptake from the soil via osmosis.

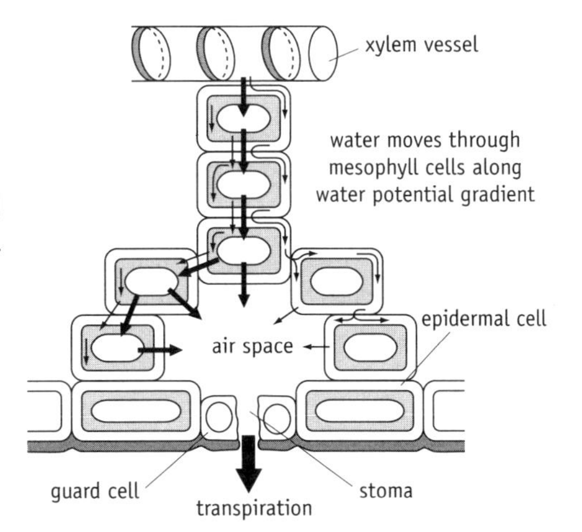

Water movement across the leaf

Make very sure that you don't confuse translocation (of organic substances in the phloem) and transpiration (of water).

✓ *Quick check 2, 3*

Evidence for movement of ions in the xylem

- **Radioactive tracers** can provide evidence that transport of mineral ions through a plant occurs in the xylem.
- Xylem and phloem in a section of the stem are separated using a wax cylinder to prevent lateral transport.
- The roots are supplied with radioactive potassium ions, $^{42}K^{+}$.

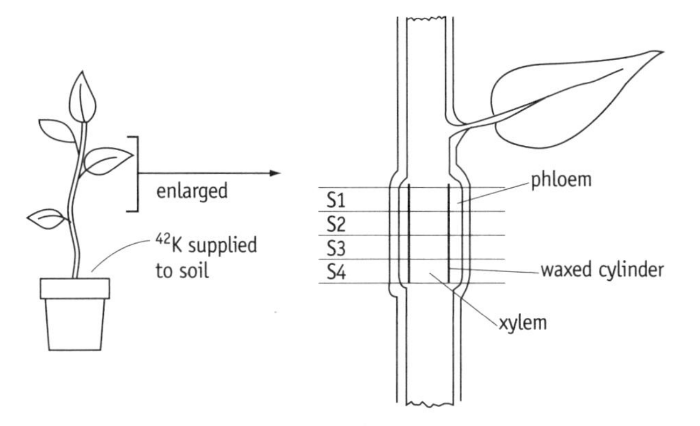

Transport of $^{42}K^{+}$ in plants

Table of results

section number	ppm* $^{42}K^{+}$ in tissue			
	stripped		unstripped	
	phloem	xylem	phloem	xylem
S1	0.9	119	49	54
S2	0.3	108	47	70
S3	0.5	112	50	67
S4	0.3	109	48	57

* ppm is a measure of concentration

- The plant is left for a few hours and the amount of radioactivity in the xylem and phloem tissues in the region of the wax cylinder is then measured and compared.
- The amount of radioactivity in the xylem is considerably greater indicating that transport of the potassium ions occurs in this tissue.
- A small amount of radioactivity in phloem tissue is due to lateral transport from the xylem in the region where the wax cylinder is not present.

Radioactive phosphate ions are often used to trace uptake from the roots to the stem – plants need phosphate for making ATP, DNA, RNA and phospholipids.

? Quick check questions

1. Briefly describe how you could show that root pressure involves active transport.
2. What is the energy source in the cohesion-tension hypothesis?
3. Explain what is meant by cohesive and adhesive forces in the transport of water in the xylem.

B

Transpiration

Transpiration is the evaporation of water from a plant's surface, particularly through the stomata. Its rate is affected by environmental factors and the structure of the plant.

> The surface of a leaf is covered by a waxy cuticle which is there to prevent water loss. It also blocks gaseous exchange – which is why stomata are needed!

Environmental factors

Light

- Stomata open in light, because guard cells have chloroplasts which react to light, leading to increased turgor of the cells.
- Opening allows carbon dioxide to enter for photosynthesis.
- More water diffuses out of the leaf – increasing the rate of transpiration.

> The worst conditions for a plant in terms of water loss is a hot day, with low humidity and a strong wind. These conditions are often found in deserts.

Temperature

- An increase in temperature gives water molecules more kinetic energy, allowing them to evaporate more easily from mesophyll cells.
- It also increases the rate of diffusion of water vapour from stomata.

Humidity

- An increase in humidity decreases the water potential gradient for the diffusion of water, decreasing the rate of transpiration.

Air movement

- Air movement removes water vapour from the leaf surface, increasing the water potential gradient and transpiration.
- In still air, water vapour builds up around the leaf reducing the water potential gradient and rate of transpiration.

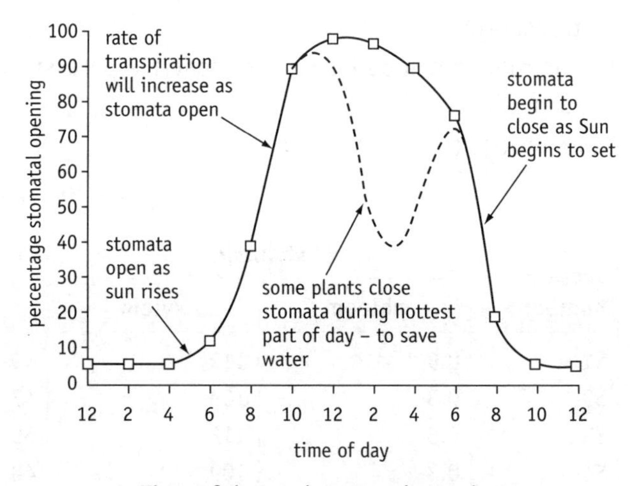

Time of day and stomatal opening

Quick check 1

Potometer

A potometer is used to measure water loss from a shoot – the stem and leaves of a plant.

The rate of water loss can be measured under different environmental conditions. This allows a **quantitative study of factors affecting transpiration**. One limitation of the potometer is that the roots are removed. There can be no root pressure, or resistance by the roots to water being pulled up the xylem in the stem.

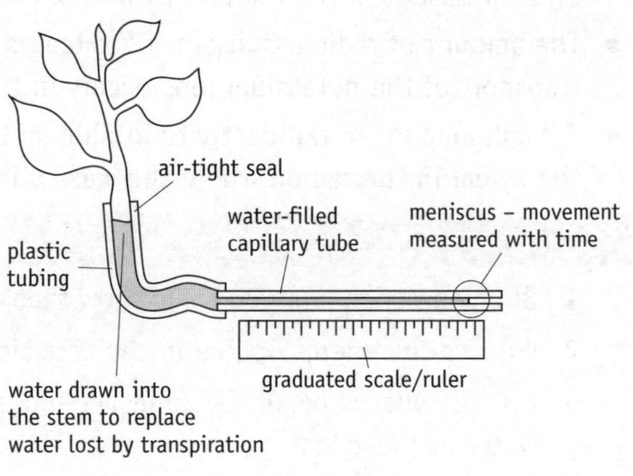

Potometer

Xerophytes

B

These are plants that live in habitats where water is in short supply.

Many have structural adaptations of leaf structure such as:

- a **thickened waxy cuticle** reducing evaporation,
- **hairs** on the leaf surface to trap a layer of air – which becomes saturated with water vapour reducing the water potential gradient for water loss,
- **curled leaves** (e.g. marram grass) to reduce the surface area for evaporation and increase the humidity in the air around the stomata, reducing transpiration.
- **reduced leaf surface area** (e.g. pine needle) over which transpiration can occur,
- **sunken stomata** (often in epidermal pits), which become saturated with water vapour reducing the water potential gradient for water loss.

Section through a leaf of marram grass to show rolled up structure to reduce transpiration

Plants which live in habitats where water supply is adequate are known as mesophytes.

The uptake of water is promoted by adaptations such as:

- a very deep and extensive root system, to reach water deep in soil,
- some very shallow roots, to absorb dew that forms on soil surface at night,
- accumulation of high concentrations of solutes in root (and other) cells – giving them a very low water potential – and making a very steep water potential gradient for the uptake of any water present from the soil.

Quick check 2, 3

Quick check questions

1. Explain why the rate of transpiration is high on a warm breezy day.
2. What are xerophytes?
3. Describe and explain how two adaptations of leaf structure can reduce the rate of transpiration.

Ⓢ Always use water potentials, rather than osmotic or solute potential or other terms. Water diffuses from a solution with a higher (less negative) to one with a lower (more negative) water potential.

B

HB

Principles of homeostasis and regulation of blood sugar

Homeostatic mechanisms keep the body in a state of **equilibrium**; maintaining a **constant internal environment** in a **changing external environment**. Homeostasis maintains **optimum conditions for enzyme activity**, in terms of factors like temperature, pH, substrate and product concentrations. **Separate mechanisms** control departures in different directions from the optimum – giving greater control.

Negative feedback

Negative feedback is a control mechanism where movement away from the normal value of something produces a response that returns it to its normal value.

- For example, the product of a series of enzyme reactions can inhibit an enzyme in the series, reducing its own production.
- The result is that the concentration of the product rises and falls slightly about a certain (optimum) concentration.

Regulation of blood sugar

The body maintains an optimum concentration of glucose in the blood (blood sugar), to supply cells with glucose for respiration. Two **hormones** control departures from normal blood sugar concentrations, **insulin** and **glucagon** – secreted by the **pancreas**.

Ⓢ Meals with lots of glucose in them quickly raise blood sugar levels. Starch-rich meals have to be digested to glucose and this takes time – so blood sugar rises after a time and slowly. This is better for diabetics.

Blood glucose concentration rises above normal

- This happens after a meal, when glucose is absorbed in the small intestine into the blood.
- Cells in the **pancreas** respond by **secreting insulin** into the bloodstream.
- **Liver** cells have **specific membrane receptors** for insulin; proteins with a tertiary structure/3D shape and receptor site which only insulin fits into.
- Binding of insulin causes glucose channels (carrier proteins) to open, allowing more glucose into cells from the blood (by facilitated diffusion).
- This returns blood sugar levels towards normal.
- Increased uptake of glucose provides a higher substrate concentration for **enzymes** inside liver cells that convert glucose into **glycogen**, an insoluble storage carbohydrate.
- As blood glucose levels return to normal, insulin secretion is reduced. The secretion of insulin eventually leads to a reduction in its secretion – **negative feedback**.

See if you can explain the control of insulin release as an example of homeostasis and negative feedback.

Quick check 1, 2

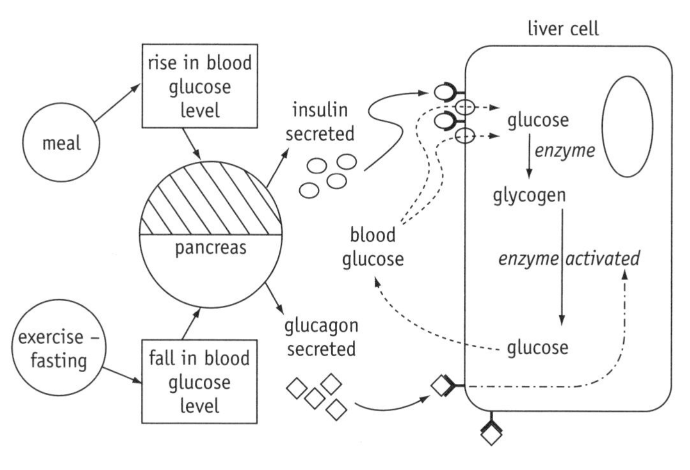

Control of blood sugar concentration

Blood glucose concentration falls below normal

- This can happen during exercise, or when someone hasn't eaten for some time.
- Cells in the pancreas respond by secreting **glucagon** into the blood.
- **Liver** cells have **specific membrane receptors** for glucagon.
- Binding leads to switching on of **phosphorylase enzymes** which break down glycogen into glucose.
- Glucose diffuses into the blood returning blood glucose towards normal.
- The fall in blood glucose concentration causes a reduction in the release of glucagon – **negative feedback**.
- The **opposing actions of insulin and glucagon** keep blood sugar from fluctuating too much about the optimum concentration.

Ⓢ Increases in substrate concentration increase the rate of enzyme reactions. Proteins (including enzymes) have specific tertiary structures that allow them to bind to specific substances with shapes that 'fit' the protein.

Don't write that insulin or glucagon react with, or act on, glucose or glycogen – they don't! These hormones affect liver cells!

Diabetes

Early onset diabetes is due to death of insulin-secreting cells and is controlled by:

- injecting insulin at regular times,
- eating food regularly and avoiding foods rich in simple sugars – to avoid sudden rises in blood sugar.

Late onset diabetes is due to a loss of sensitivity of target cells to insulin, because of excessive release of insulin over many years. It is controlled by:

- cutting down on carbohydrate in the diet – to reduce the amount of insulin released into the blood and allow target cells to recover,
- increasing exercise, to increase glucose demand by muscle cells.

✓ Quick check 3

Quick check questions

1 Explain why insulin levels in the blood rise and fall at various times during a day.

2. Explain how insulin returns blood sugar levels to normal after a sugar-rich meal.

3 Explain how each of the following is involved in the regulation of blood sugar by insulin and glucagon; (**a**) specific membrane receptors (**b**) enzymes.

B

HB

Regulation of body temperature

Regulation of body temperature is important, because **enzymes** work fastest at their **optimum temperature** and this affects the metabolic rate of an organism. **Thermoreceptors** in **blood vessels** in the **hypothalamus** (in the brain) detect rises or falls in blood temperature. Thermoreceptors in the **skin** detect changes in environmental temperatures.

Reptiles are ectotherms

Reptiles:

- **do not control the temperature of their bodies within narrow limits,**
- rely on heat from the environment to raise body temperature, rather than/as well as heat from metabolism.

Reptiles control their body temperature by **behaviour** by:

- basking in sunshine, to absorb heat,
- finding shade or cool places to lose heat,
- moving – which involves muscle contraction, respiration and the release of heat.

Mammals are endotherms

Mammals:

- **maintain a constant body temperature**, independent of the temperature of the external environment,
- rely mainly on heat from metabolism, most from respiration.
- control loss of heat released from respiration.

Heat is lost by radiation, convection, or conduction to cold surfaces.

If blood or environmental temperature starts to rise:

- nerve impulses travel to a **heat loss centre** in the **hypothalamus**, the coordinator,
- which sends **nerve impulses to effectors** (mainly in the skin),
- **responses** increase heat loss and lower the temperature of the blood.

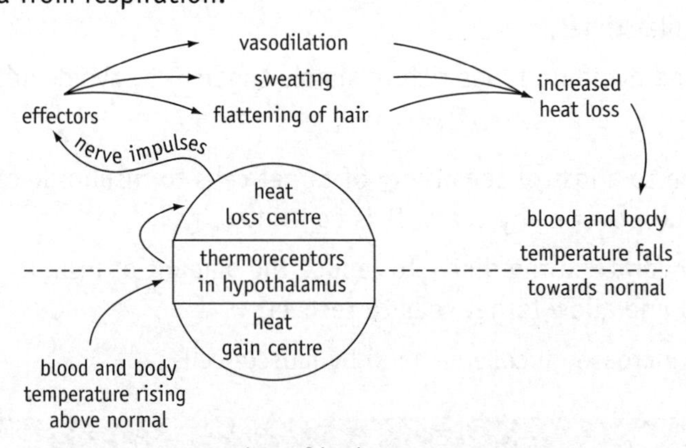

Lowering of body temperature

Vasodilation

- Circular muscles around arterioles in the skin relax, allowing more blood to flow near the surface of the skin.

Blood vessels – such as capillaries – do not move during vasodilation (or vasoconstriction)!

B

HB

Sweating

- Sweat glands release more sweat which evaporates from the skin,
- getting its heat of vaporisation from the skin (and blood).

Flattening of hair

- Muscles at the base of hairs relax, allowing the hair to lie flat,
- reducing the insulating air layer next to the skin.

Quick check 1

Behaviour

Humans can take off clothing, or find shade from the Sun.

All of these responses can increase heat loss by radiation, convection (warm air rising off the body), or conduction (if you are touching a cool surface).

If blood or environmental temperature starts to fall;

- nerve impulses travel to the **heat gain centre** in the hypothalamus,
- **responses** reduce heat loss (and sometimes cause more heat production).

Vasoconstriction

- Less blood flows near the surface of the skin.

Less sweat is released.

Erection of hair

- Muscles at the base of hairs contract and hairs stand-up,
- increasing the insulating air layer - reducing heat loss.

Shivering

- Muscles contract in spasms, producing more heat from respiration.

Metabolic rate

- The adrenal glands secrete the **hormone noradrenaline**,
- increasing respiration and heat production.

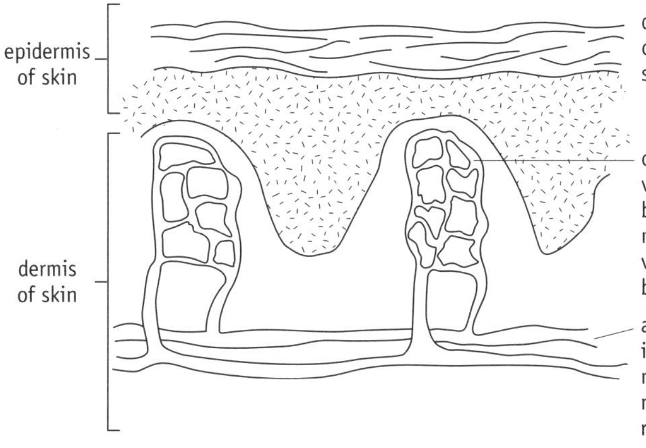

Vasodilation and vasoconstriction

Negative feedback and inhibition

- An active heat loss centre sends inhibitory nerve impulses to the heat gain centre.
- As heat loss increases, body temperature falls until it drops below normal.
- The heat gain centre then becomes active, inhibits the heat loss centre and makes the body temperature rise again.
- This is control by negative feedback.

During exercise, muscles need more energy from respiration. More heat is produced in muscles. This heat is carried away by the blood to the skin.

Quick check 2, 3

If you have to write about the response of the body to a change in temperature, think about the sequence – stimulus ⇒ receptor ⇒ coordinator ⇒ effector ⇒ response.

Quick check questions?

1 Explain how body temperature is kept constant during exercise.

2 Drinking an ice-cold drink can lower the temperature of blood flowing in vessels near to the stomach. Suggest how an ice-cold drink might cause a temporary reduction in heat loss from the skin.

3 People with hypothermia have a body temperature below normal. They become less and less active, their body temperature continues to fall and can eventually become unconscious and die. Suggest why the progressive fall in body temperature takes place.

B

Methods of removing nitrogenous waste

Waste products of metabolism are toxic/harmful if they accumulate and are excreted. Animals can not store amino acids or protein.

- **Surplus amino acids** (not needed immediately for protein synthesis) have their amine group removed, leaving an **organic acid.**
- The organic acids are used as **respiratory substrates**, to yield energy.
- The amine group is used to make **ammonia** which is **toxic** if it accumulates in the body.
- Ammonia is **excreted** directly, or used to make another substance which is then excreted.

Ⓢ Make sure you know the basic chemical structures of the important types of biological molecules from Module 1.

Fish

- Fish **excrete** highly soluble **ammonia** into the water, mainly from the **gills.**
- Ammonia diffuses from the blood in the capillaries in gill lamellae into the large volumes of water passing over the gills.
- The water current over the gills rapidly carries ammonia away and maintains a **steep diffusion gradient** for ammonia.
- The ammonia is then **diluted away** in the water of the river/lake/sea.

Mammals can only tolerate very low concentrations of ammonia in their blood – a maximum of about 0.2 mg per litre of blood.

Insects and birds

- Birds and most insects are terrestrial (land-living) and need to prevent excessive water loss.
- They do not have an aqueous environment to excrete ammonia into.
- They make **uric acid**, which is insoluble in water.
- This is much **less toxic** than ammonia and can be stored in their bodies until excreted (and in their eggs!)
- It is excreted as small crystals in a very small volume of water – as a white paste.

Some insects living in desert conditions excrete almost totally dry crystals of uric acid – to save water!

Mammals

- Most mammals are terrestrial and all need to control water loss.
- They produce **urea.**
- Each molecule carries two amine groups away and is **less toxic** and soluble than ammonia.
- It can be tolerated in the body in reasonable amounts and excreted in a small volume of water.

Mammals normally have up to 0.4 mg per litre of urea in their blood.

✓ *Quick check 1, 2*

Deamination and urea production

In mammals, proteins are digested in the gut and the resulting amino acids are absorbed in the small intestine into the blood. The blood carries the amino acids to the liver.

- The diet usually contains a surplus of amino acids but not the correct amount or types of all amino acids.
- Amino acids needed for protein synthesis are used.
- Surplus amino acids can be converted into other types that are needed.
- Remaining surpluses can not be stored and are **deaminated** in liver cells.
- The **amine group is removed** from the amino acid, leaving an **organic acid.**
- The organic acid can be used as a respiratory substrate, to provide energy.
- The amine group is used to make **urea.**
- Urea leaves liver cells and goes into the blood, is removed from the blood by the kidneys and excreted in urine.

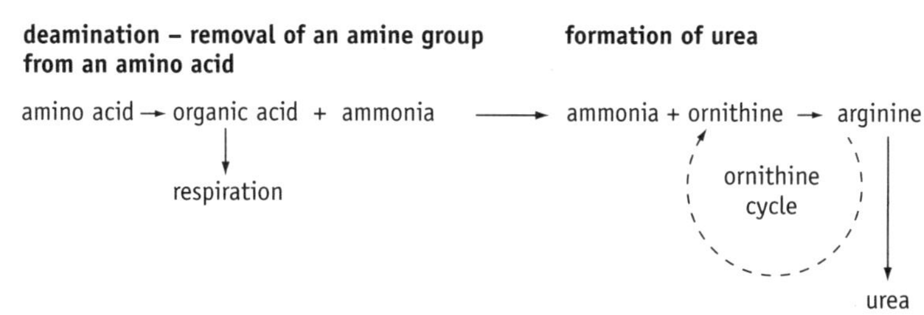

Production of urea

A carnivore's diet will contain a lot of protein in the meat it eats – so this will be a major energy source (along with fats).

Quick check 3

Don't learn lots of unnecessary details about the ornithine cycle!

Quick check questions

1. Explain why fish excrete their nitrogenous waste as ammonia across the gills but mammals excrete urea in urine.
2. Some insects are found in the driest deserts. Explain how their method of nitrogenous waste excretion helps them to survive.
3. Explain why, where and how urea is formed.

B

Kidney 1 – Urine production

The kidneys remove urea from the blood and form urine. They help maintain the water, pH and mineral ion balances of the blood. The kidney consists of vast numbers of nephrons.

Ⓢ Make sure you know the general pattern of blood circulation. How does a molecule of urea get from the liver to the kidney in the blood?

Nephrons and associated blood vessels

- Each **nephron** consists of a **renal capsule, proximal convoluted tubule, Loop of Henle** and **distal convoluted tubule.**
- Nephrons are connected to **collecting ducts.**
- The renal capsule is also called the Bowman's capsule.

Ultrafiltration

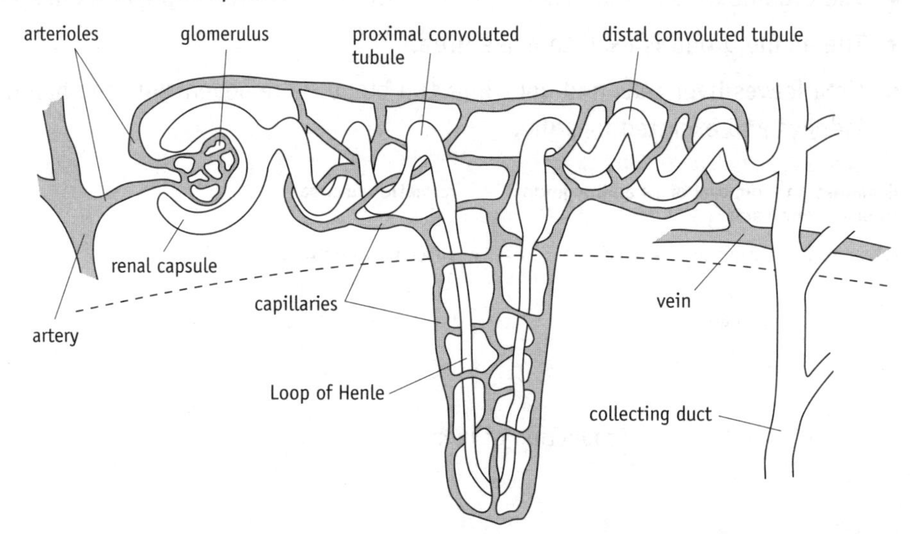

Nephron and blood supply

- Blood enters the kidney under high pressure through the renal artery.
- This branches into short arterioles with **glomeruli** – where blood pressure is high.
- Glomerular capillaries have endothelial cells with 'pores', allowing substances to leave the blood without crossing cell membranes.
- There are 'pores', or 'slits', between cells (podocytes) of the **renal capsule.**
- A basal lamina surrounds the capillaries – a mesh of fibrous connective tissue protein.
- A lot of blood plasma is forced through the 'pores' and lamina into the tubule by the **high blood/hydrostatic pressure**, forming the **filtrate – ultrafiltration.**
- The pores and lamina act as a filter, **letting through** small substances like water and dissolved urea, glucose and mineral ions but **not** blood cells and large blood proteins.

Ⓢ Blood consists of blood plasma and suspended blood cells. Plasma contains dissolved glucose, mineral ions, amino acids, urea and proteins.

✓ Quick check 1

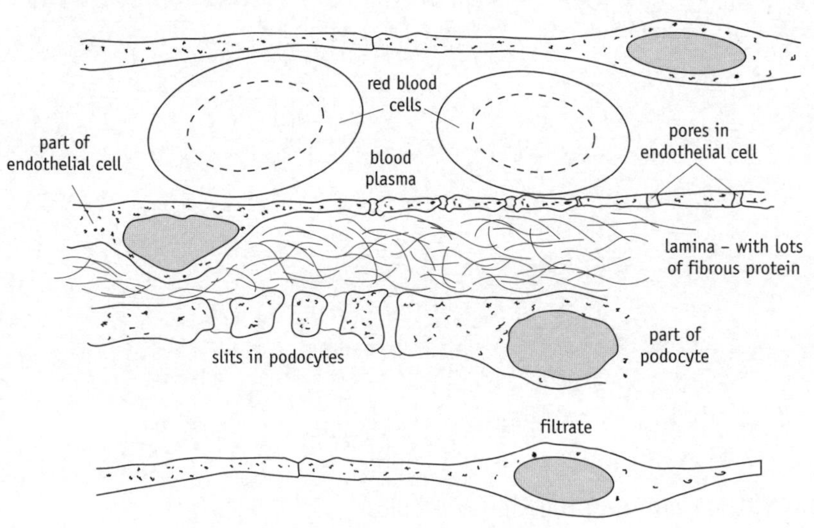

Ultrafiltration and structure in the renal capsule

B

Selective reabsorption of useful substances

Useful substances in the filtrate are **reabsorbed into the blood** along the tubule. Reabsorption of water is vital, to avoid dehydration.

Filtrate leaving the renal capsule enters the next part of the tubule.

Proximal convoluted tubule

- **Glucose, Na^+ and K^+** ions are reabsorbed from the filtrate by **active transport/uptake**, using **specific carrier proteins** in cell membranes of tubule cells.
- No glucose is left in the filtrate or excreted with urine.
- Active transport uses a lot of ATP/energy from respiration.
- Blood plasma in capillaries around the proximal tubule has a very low water potential compared with the filtrate (due to dissolved blood proteins and reabsorbed ions and glucose).
- This causes water to be reabsorbed into the blood by **osmosis**.
- Urea is not reabsorbed.

Ⓢ Specific and active reabsorption of substances from the filtrate involves carrier proteins in the membranes of tubule cells. These are specific because of their specific tertiary structures.

✓ Quick check 2

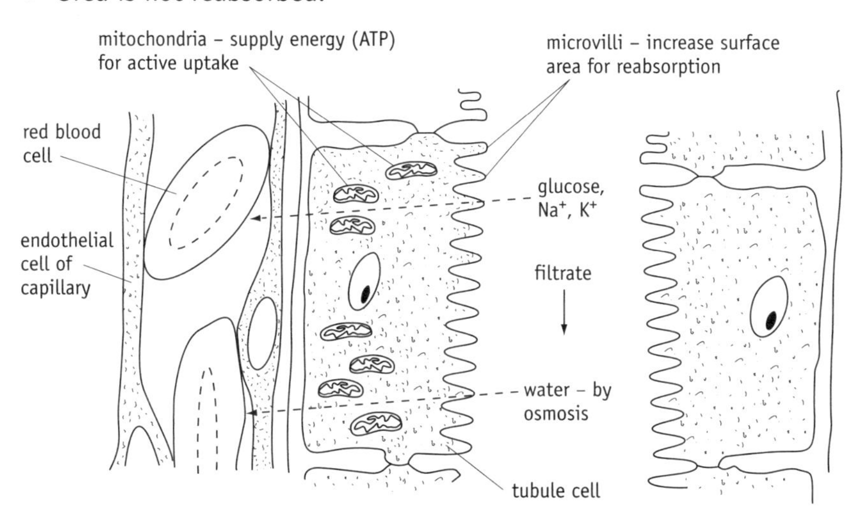

Reabsorption in the kidney tubule

Loop of Henle

More ions and water are reabsorbed.

Distal convoluted tubule

More ions and water are reabsorbed, producing **urine**, a concentrated solution of urea.

Collecting duct

This collects urine from many tubules.

- More water can be removed from the urine in the collecting duct.
- The amount of water lost in the urine is controlled by the amount of water reabsorbed by the distal tubule and the collecting duct (see next section).

Make sure you are answering the question set and don't write everything you know about the kidney! If the question asks about reabsorption of glucose or ions, don't write about water!

✓ Quick check 3

Quick check questions

1. Describe and explain the differences between blood plasma and the filtrate entering the kidney tubule.
2. Patients' urine is often tested for the presence of either glucose or protein. Suggest **one** reason for the presence of either of these in a urine sample.
3. Explain how urine is formed from the filtrate entering the kidney tubule.

B

Kidney 2 - Water balance

The water potential of blood plasma is kept constant, which keeps the water potential of tissue fluid and cell contents the same – leading to no net gain, or loss, of water by osmosis.

- If the body has **too much water**, less water is reabsorbed by the distal tubule and collecting duct.
- A **greater volume** of **dilute** urine is produced.
- If the body has **too little water**, more water is reabsorbed.
- A **smaller volume** of **concentrated** urine is produced.

Removing water from the filtrate/urine in the distal tubule and collecting duct is difficult, because it has a very negative water potential. A lot of water has already been reabsorbed by the proximal tubule and Loop of Henle.

Ⓢ Make sure you understand water potentials. Pure water has a water potential of 0. The more substances there are dissolved in it, the lower the water potential. There is a net flow of water by osmosis from a higher to a lower water potential.

Loop of Henle

To reabsorb more water by osmosis, the surrounding tissue (fluid) has a lower water potential than the filtrate/urine.

- The cells of the **Loop of Henle** carry out **active transport** of **chloride ions from the filtrate into the surrounding tissue** (fluid) – **which also surrounds the distal tubule and collecting duct.**
- Sodium ions 'follow' the chloride ions. **In effect, sodium chloride is pumped into the surrounding tissue** (fluid), **giving it a more negative water potential than the filtrate/urine.**
- Water can be reabsorbed by **osmosis** from the filtrate into the surrounding tissue and then into the blood.

✓ Quick check 1

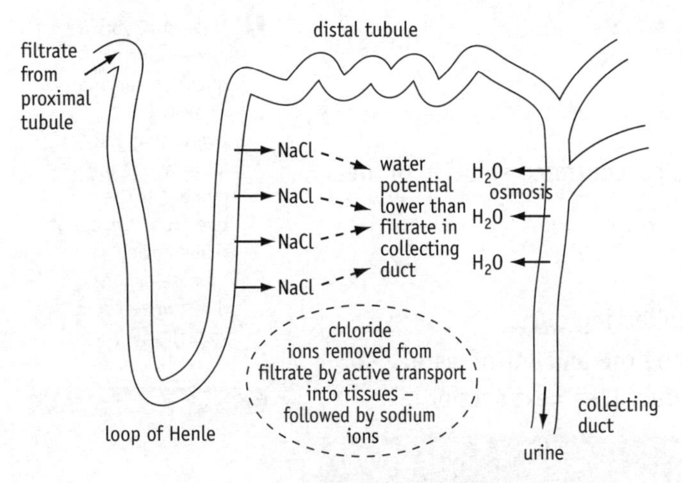

Water reabsorption in the kidney

The countercurrent multiplier hypothesis

The longer the loop of Henle, the more water can be reabsorbed from the filtrate, reducing water loss. Desert-living animals usually have very long loops of Henle.

- The **loop of Henle** consists of a **descending limb** and an **ascending limb**, with **filtrate flowing in opposite directions** – a **countercurrent.**
- The **ascending limb** is **impermeable to water** but **actively transports sodium chloride out of the filtrate.**
- Filtrate flowing up the ascending limb **becomes less concentrated** as sodium chloride is actively transported out into the surrounding tissues.
- The size of gradient that sodium chloride is transported against stays the same, because **filtrate flowing down the descending limb loses water by osmosis** into the surrounding tissue and **becomes more concentrated.**
- The **descending limb** is **permeable to water** but **impermeable to** mineral ions – including **sodium chloride.**
- The **water is taken up by specialised capillaries**, the **vasa recta.**
- The longer the loop of Henle, the more concentrated the filtrate is at the base of the loop and the more water is reabsorbed.

Most people find this hard to understand! Just try to remember the basics here.

✓ Quick check 2

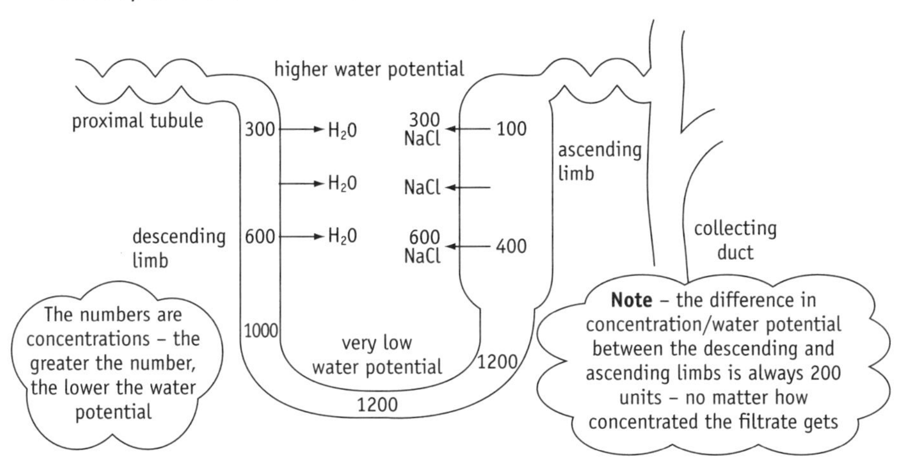

Countercurrent multiplier

Make sure you are answering the question set! If you are asked about how water is reabsorbed, that's the last section. If you're asked about how the amount of water reabsorbed is controlled, that's this section.

Control of the amount of water reabsorption

If the body is losing too much water, the blood water potential becomes slightly more negative.

- This stimulus is detected by receptors in the blood vessels of the hypothalamus, in the brain.
- The response is the release of the hormone **ADH** from the pituitary gland, into the blood.
- ADH makes cells of the collecting duct and distal tubule more permeable to water and more water is reabsorbed.
- If the body has too much water, less/no ADH is released and less water reabsorbed.

✓ Quick check 3

? Quick check questions

1. Explain how water is reabsorbed from filtrate in the collecting duct.
2. Explain why a long loop of Henle can lead to less water loss.
3. During prolonged exercise a person sweats a great deal and does not drink. Suggest how the body would react to this loss of water.

B

Gaseous exchange surfaces

Gaseous exchange between an organism and its environment relies on **diffusion**.
Gaseous exchange surfaces have:

- **large surface areas** to increase the rate of diffusion;
- **thin surface** layers (epithelia) – providing a short diffusion pathway;
- **steep diffusion gradients** for oxygen and carbon dioxide.

They may also have,

- extensive blood supplies,
- and a ventilation mechanism which **maintains** steep diffusion gradients.

The **larger an organism, the smaller the surface area:volume ratio.**

- In large animals blood transport systems carry gases between exchange surfaces and tissues.

Adaptations have evolved which maintain adequate exchange, including:

- changes in body shape, to increase the surface area, e.g. flattening of the body.
- **internal respiratory systems with large surface areas relative to the volumes of organisms.**

Fick's law reflects these features.

$$\textbf{diffusion rate} \propto \frac{\textbf{surface area} \times \textbf{difference in concentration}}{\textbf{thickness of exchange surface}}$$

> Remember – **small** organisms have a **large** surface to volume ratio and vice versa for large organisms.

> Ⓢ See the section in Module 1 about how substances cross plasma membranes.

Protoctistans

- Single-celled organisms have a **large surface area to volume ratio.**
- They have short diffusion pathways to all parts of the cell.
- Steep diffusion gradients are maintained by respiration.

amoeba – a protoctistan

O_2 concentration lower than environment, because O_2 constantly used in respiration

unicellular – no part of the cytoplasm is more than a fraction of 1 mm from the outside

O_2 diffuses in

CO_2 diffuses out

Gaseous exchange in an amoeba

✓ Quick check 1

Gaseous exchange in plants

A waterproof waxy cuticle covers leaves, allowing little gaseous exchange.

- Gases diffuse in and out through stomata.
- Many stomata and thin leaves leads to short diffusion pathways.

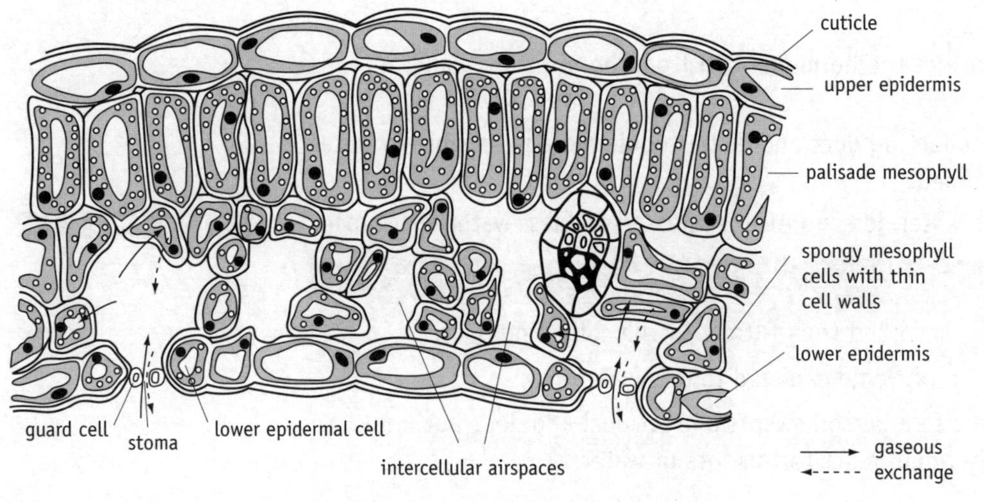

Gaseous exchange in leaves

> Ⓢ Plant cells respire all the time! Only respiration takes place in leaf cells at night – so oxygen enters the leaf and carbon dioxide leaves. In daylight, the cells respire **and** carry out photosynthesis. When the rate of photosynthesis is greater than the rate of respiration, there is a net uptake of carbon dioxide and loss of oxygen.

> You might get a question that talks about gas exchange in an organism you've never heard of! Don't panic – look for the basic properties of exchange surfaces – they must be there!

B

- Respiration and photosynthesis maintain diffusion gradients by using and producing oxygen and carbon dioxide.
- Spongy mesophyll cells lining air spaces provide a large surface area.
- Gases diffuse rapidly through intercellular air spaces and thin cell walls of mesophyll cells.
- Gases diffuse quickly across thin cell walls and membranes of cells.

Gaseous exchange in insects

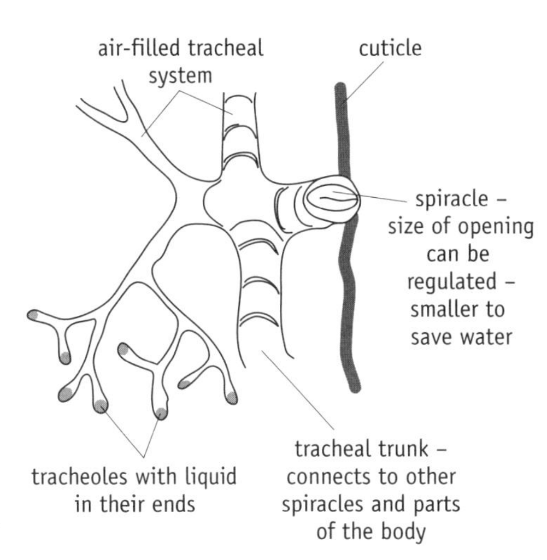

Tracheal system of insects

Insects are covered by a waterproof **cuticle**, with a layer of wax on its surface.

- **Spiracles** (openings on the abdomen) connect to air passages – **trachea**.
- These branch into small, dead-end **tracheoles** – where gaseous exchange takes place – all cells are very close to a tracheole.
- Oxygen and carbon dioxide diffuse rapidly through air in the tracheal system.
- Tracheoles have thin walls and together have a large surface area relative to the volume of the insect.

✓ Quick check 2

Gaseous exchange in fish

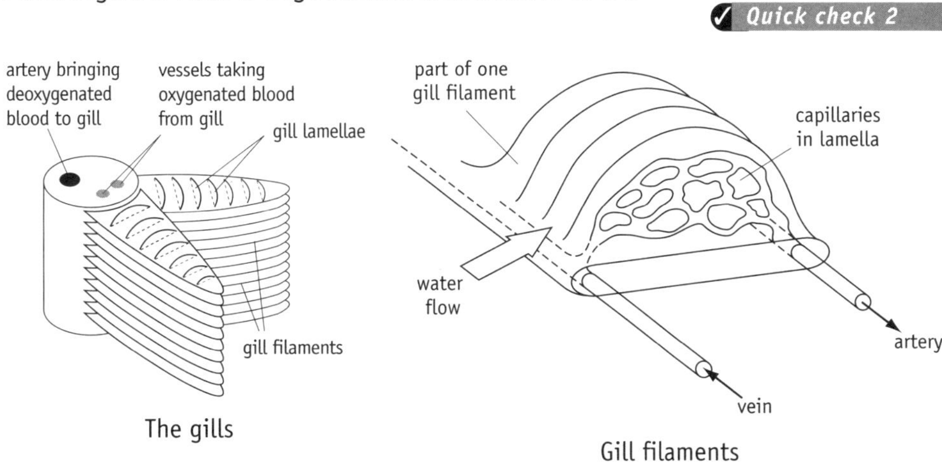

The gills

Gill filaments

In fish, gaseous exchange occurs over the surface of the **gills**.

- A gill has two rows of **gill filaments** with **lamellae** (thin plates), providing **a large surface area** and **short diffusion pathways**.
- The lamellae have an abundant supply of blood capillaries.
- A thin barrier of two cell layers (epithelial and endothelial) provides a **short diffusion pathway** between the blood and water.
- A ventilation mechanism makes water flow over the gills in the opposite direction to blood (countercurrent system), bringing oxygen and removing carbon dioxide and **maintaining a steep diffusion gradient**.

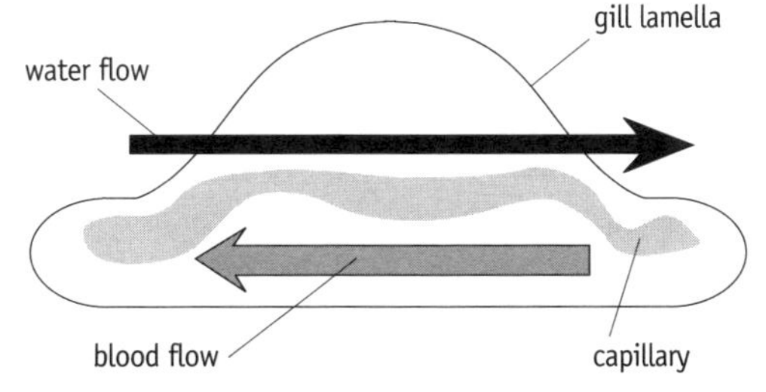

Countercurrent system

Air goes in and out of our lungs but water flows in one direction over the gills. This is because water is too dense to easily breathe in and out – and contains less oxygen than air, so more has to pass over the exchange surface.

✓ Quick check 3

Quick check questions

1. Explain why protoctistans do not need special gas exchange surfaces.
2. Describe the similarities between the gaseous exchange systems of plants and insects.
3. Explain how the gills of a fish are adapted for gaseous exchange.

B

Limiting water loss

Terrestrial (land-living) organisms have adaptations to obtain enough water and control water loss, to avoid desiccation/dehydration.

Small desert mammals

These animals face special problems because of:

- shortages/absence of water to drink,
- food which is often quite dry itself,
- high environmental temperatures,
- and low air humidity that increase evaporation.

Obtaining water

Preformed water is obtained from:

- **drinking** water from water-holes, streams (after rain), or dew,
- water in **food**.

Metabolic water:

- comes from **condensation reactions** in metabolism.
- **Respiration** is the main source.
- Using fat as a respiratory substrate yields more water than carbohydrate.

Ⓢ In aerobic respiration, oxygen is the final acceptor at the end of the electron transport chain. Oxygen reacts with an electron and a proton (H^+) to form water.

Minimisation of water losses

Water can be lost when breathing, from the skin and when excreting.

- **Breathing** – **respiratory surfaces** of the lungs are wet and water evaporates into air that is exhaled.
- Air is moistened as it is inhaled through the nose–increasing air humidity and reducing evaporation from lung surfaces.

Skin losses are reduced by having:

- dead, keratinised cells over the surface,
- oily secretions that waterproof the skin,
- by not sweating.

Excretory losses are reduced by having kidneys that produce very concentrated urine.

Behavioural adaptations

- Many desert mammals stay in burrows when temperatures and thus evaporation are greatest.
- Some seek shade under trees or rocks.

Water budget of desert mammal for one month	
Water gains/cm^3	**Water losses/cm^3**
oxidation water from respiration	urine 13.5
oxidation water from carbohydrate in seeds 54	faeces 2.6
from water in seeds 6	evaporation from lungs and skin 43.9
TOTAL 60	60

✓ Quick check 1

Mesophytic plants

These are terrestrial plants that live where there is a sufficient supply of water.

- Leaves are covered by a **waxy cuticle**, to reduce water loss by evaporation.
- The cuticle prevents gaseous exchange, so there have to be stomata.
- The upper surface of the leaf gets the hottest and has the thickest cuticle.
- Stomata may close during the hottest part of the day to conserve water.

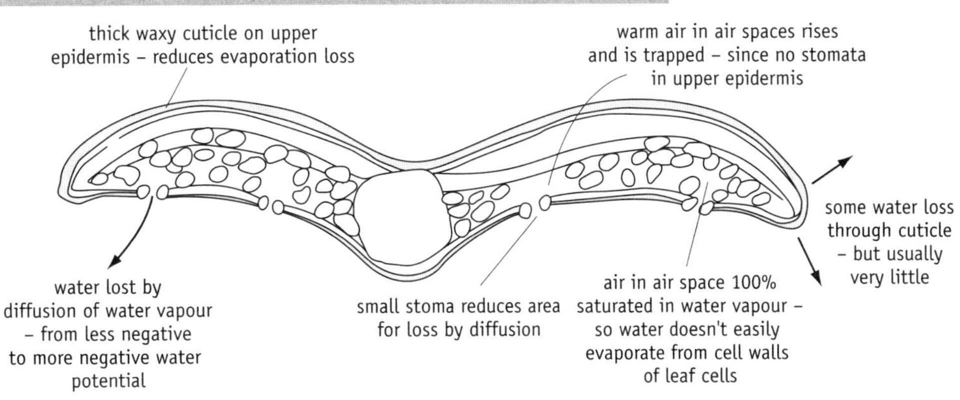

Water loss and conservation in a mesophytic leaf

Insects

Insects have an **exoskeleton** made of **chitin**. This has a **wax** layer on it which makes it waterproof.

- The wax is thickest in species that live in dry environments.
- Water is lost by diffusion from spiracles – openings to the tracheal system.
- The largest tracheae are lined with exoskeleton (or rings of this material) to prevent their collapse and this also reduces water loss.

When an insect is at rest:

- not much oxygen is required for respiration – spiracles almost closed to reduce water loss,
- tracheoles partly filled with fluid reduce the surface area in contact with air in the tracheal system,
- air in the tracheal system is almost saturated with water vapour, reducing evaporation of water from the surfaces of tracheoles.

When an insect is active:

- its rate of respiration and the concentration of carbondioxide increases,
- causing spiracles to open and increasing gaseous exchange and water loss,
- liquid is absorbed from the tracheoles, exposing a greater gas exchange surface but increasing water loss.

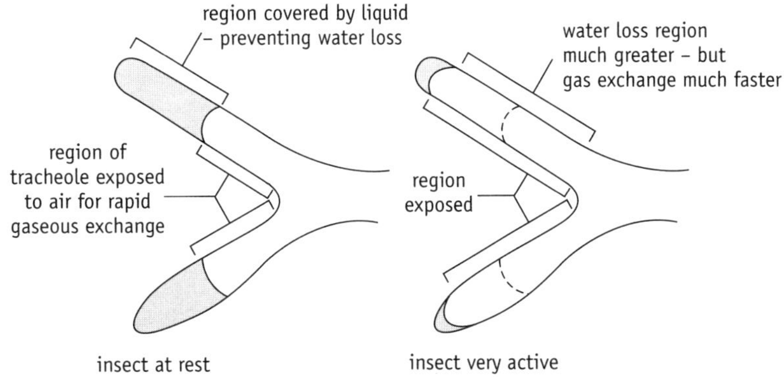

Tracheole filling

Quick check 2

Oxygen and carbon dioxide could diffuse to and from the atmosphere through cells. This would involve diffusion through water which is much, much slower than diffusion through air. Having air-filled passages between the atmosphere and gas exchange surfaces gives much faster diffusion.

Quick check 3

Ⓢ When an animal is active (or 'exercising'), its muscles need more energy. This comes from increased respiration, which requires more glucose and more oxygen.

? Quick check questions

1. A small desert mammal was never seen to drink and only fed on dry seeds. Explain how it was able to survive.
2. Explain how a leaf conserves water but still carries out gas exchange.
3. A desert-living insect was seen to catch prey from ambush; it hid under a stone and then pounced on any suitable prey. Suggest how this style of hunting helped it to conserve water.

B

HB

Transport of respiratory gases

Cells need oxygen for respiration and need to get rid of waste carbon dioxide.

> Haemoglobin is a protein with specific binding sites for 4 oxygen molecules – shape plays a key role in binding.

Haemoglobin and hydrogencarbonate ions

Oxygen is carried in **red blood cells**, **reversibly** bound to **haemoglobin**. Most **carbon dioxide** is carried as **hydrogencarbonate ions** in **blood plasma**.

- In body tissue, carbon dioxide diffuses into capillaries and red blood cells.
- Here **carbonic anhydrase** uses carbon dioxide to make **carbonic acid**, which then dissociates into H^+ ions and hydrogencarbonate ions.
- H^+ ions would lower the pH of the blood but they are **buffered**.
- **Haemoglobin** is a **buffer** – it takes up H^+ ions – which also promotes release of oxygen from oxyhaemoglobin.

Cl^- chloride ions leave to balance HCO_3^- ions entering

HCO_3^- in blood plasma

CO_2 excreted in the lungs

carbonic anhydrase breaks down carbonic acid

in lung capillaries

$HCO_3^- + H^+ \rightarrow H_2CO_3 \rightarrow CO_2 + H_2O$

oxygen absorbed in the lungs

red blood cell

haemoglobin + oxygen → oxyhaemoglobin

in muscle capillaries

oxyhaemoglobin → haemoglobin + oxygen

(buffer)

oxygen to respiring tissue

CO_2 from respiring tissue

$CO_2 + H_2O \rightarrow H_2CO_3 \rightarrow H^+ + HCO_3^-$

carbonic anhydrase forms carbonic acid

HCO_3^- (buffer) in blood plasma

Cl^- chloride ions enter to balance HCO_3^- ions leaving

Role of haemoglobin in respiration

- **Hydrogencarbonate** is also a **buffer** in the blood plasma.
- In the lungs, carbon dioxide diffuses out of the blood – giving a lower concentration which causes **reversal** of the carbonic anhydrase reaction.

> The Bohr effect helps to match the amount of oxygen released by the blood to the rate of respiration in a tissue.

The Bohr effect

During exercise, muscles use more energy, **respiration increases**, more oxygen is used, **more carbon dioxide** is produced and **more H^+ ions**.

- **Oxygen-haemoglobin dissociation curves** show the relationship between the amount of oxygen carried by the blood and oxygen in the tissues.
- In the steep part of the dissociation curve, a small fall in oxygen in the tissues causes a lot of oxygen to dissociate from haemoglobin.
- H^+ ions promote release of oxygen from oxyhaemoglobin.
- The greater the rate of respiration in a tissue, the more carbon dioxide (and lactic acid) released, the more carbonic acid formed and H^+ ions present.

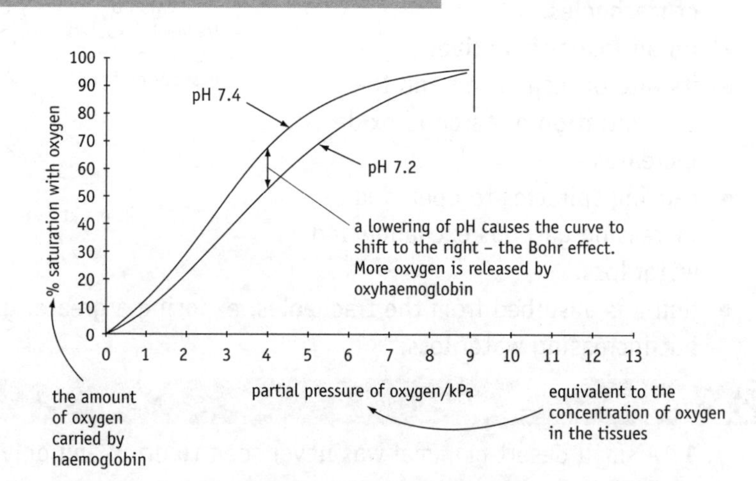

Haemoglobin dissociation curve

✓ Quick check 2

B

HB

Different types of haemoglobin

Each species that has haemoglobin has its own type of haemoglobin. The dissociation curve for a haemoglobin reflects the environment an organism lives in and how the organism lives.

- In environments with little oxygen, organisms have haemoglobin with a very high affinity for oxygen – it saturates at low oxygen pressures.
- The higher the metabolic rate of tissues (e.g. in mammals), the more oxygen is needed.
- This haemoglobin saturates at high oxygen pressures – holds a lot of oxygen.

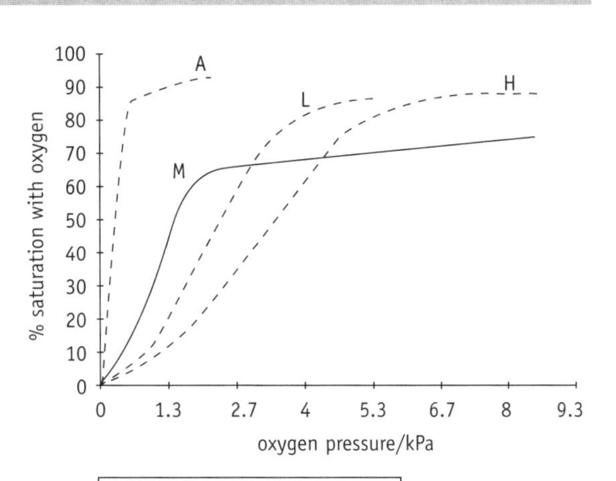

A – worm living in mud where there is very little oxygen – haemoglobin with very high affinity for oxygen
M – fish living in water where there is less oxygen than in air
L – llama living at high altitude where there is less oxygen than at sea level but more than in water
H – human living at low altitude

Different types of haemoglobin

✓ Quick check 3 HB

Fetal haemoglobin

Fetal haemoglobin is different to adult haemoglobin – it has a **greater affinity for oxygen**.

- This allows it to take oxygen from the mother's blood across the placenta.

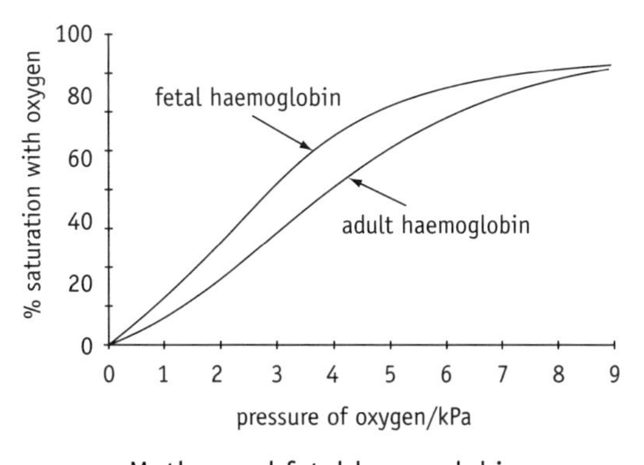

Mother and fetal haemoglobin

The mother's blood and the fetus' blood never mix – the mother's blood does not flow into the fetus!

✓ Quick check 4

? Quick check questions

1. Explain how carbon dioxide helps to release oxygen in body tissues.
2. Explain why more oxygen is released into muscle tissue during exercise.
3. A student found a bright red worm living in the mud at the bottom of a pond. Suggest why the worm is red and what advantage this gives.
4. Explain why oxygen crosses the placenta into the fetus's blood.

B

HB

Digestion and absorption of food

Food contains **biological molecules – carbohydrates, proteins and lipids.** Some are **large polymers**; often **insoluble** and **too large to absorb.**

- Large polymers are digested into small, **soluble** molecules which can be **absorbed** in the gut.
- Digestion in mammals is physical (chewing) and chemical (enzymes).
- **Digestive enzymes** in the gut **hydrolyse** large food molecules.

Ⓢ In Module 1 you studied the formation of polymers through condensation reactions. You also learnt that they can be broken down by hydrolysis.

Digestion of starch

Starch is a large, insoluble polymer of glucose.

- Starch is **hydrolysed** to **maltose** by **salivary amylase** in the mouth and **pancreatic amylase** in the small intestine.
- Maltose is hydrolysed by **maltase** in **microvilli** of intestinal epithelial cells.
- Glucose is taken up by epithelial cells by **active transport**.
- The intestinal cells secrete glucose into blood capillaries.

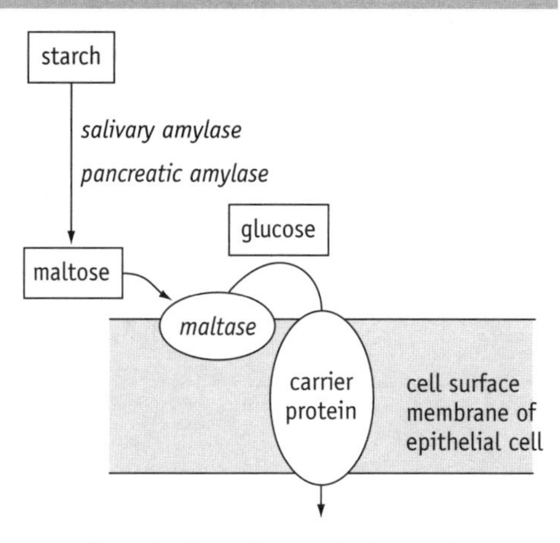

Starch digestion and absorption

Chewing (mastication) breaks the food into smaller particles increasing the surface area for enzymes to act on, so that digestion is faster.

Quick check 1

Digestion of proteins

Proteins are polymers of amino acids.

- **Gastric juice** (stomach) contains **pepsinogen** and **hydrochloric acid.**
- **Pepsinogen** converts to **pepsin** in the optimum acidic pH of 1.8.
- **Pepsin** is an **endopeptidase** – it **hydrolyses** peptide bonds in the middle of **proteins/polypeptides**, producing **shorter polypeptides.**
- Pancreatic juice released into the small intestine contains **trypsinogen.**
- This is converted to **trypsin** (**endopeptidase**) by **enterokinase**; produced by epithelial cells of the small intestine.
- **Trypsin** itself **activates trypsinogen.**
- **Exopeptidases** in **pancreatic juice** and on **epithelial cells** remove short peptides and amino acids from the ends of polypeptides.
- **Peptidases** in epithelial cells – hydrolyse dipeptides to amino acids.

Ⓢ The acidic pH removes a short chain of amino acids from one end of pepsinogen. This leads to a shape change in the rest of the molecule, to give the tertiary structure of pepsin.

Digestion of protein

Quick check 2

B

HB

Digestion of lipids

Triglycerides are digested to fatty acids and glycerol.

- **Bile** produced by the liver, is stored in the gall bladder and enters the small intestine by the bile duct.
- It contains **sodium hydrogen carbonate** and **bile salts**.
- It **neutralises** acidic chyme from the stomach and provides a slightly alkaline, optimum pH for digestive enzymes.
- Bile salts **emulsify** lipids, forming small fat droplets.
- This **increases the surface area** for the action of **pancreatic lipase**.
- **Lipase** hydrolyses lipids into **fatty acids and glycerol** which then;
- associate with bile salts and cholesterol to form micelles,
- and as lipids are then able to diffuse cross the membrane of intestinal epithelial cells,
- where they are re-synthesised into triglycerides and used to make phospholipids.
- These combine with proteins to form **chylomicrons**, which are secreted into **lacteals** of the lymphatic system.

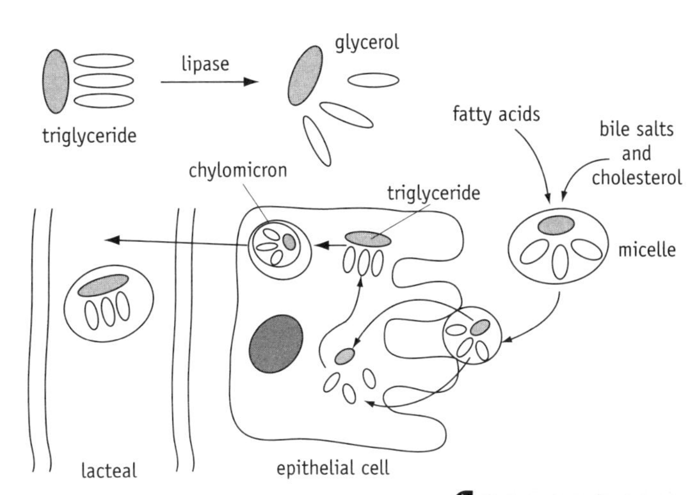

Digestion of lipids

Digestive enzymes work together to speed up digestion. Endopeptidases produce lots of shorter polypeptides with lots of 'ends' for exopeptidases to work on – this produces lots of dipeptides for dipeptidases to work on. Can you see a similar pattern for digestion of starch?

✓ Quick check 3

Digestion of cellulose

The plant food of herbivores contains a lot of cellulose and relatively little protein. Mammals do not produce **cellulase** to hydrolyse cellulose to glucose.

- Some **bacteria** and **protoctists** produce **cellulase**.
- **Ruminants** have **mutualistic** bacteria and protoctists in their **rumen** – a large chamber formed by a modified part of the gut.
- The **microorganisms get** food, warmth, water and anaerobic conditions.
- The **ruminant gets** cellulose digested and supplies of protein and vitamins.

The rumen is a **fermenter**, where anaerobic **bacteria**:

- **digest cellulose** to glucose,
- **make proteins**, using ammonia excreted by other microorganisms,
- **make vitamins**, such as vitamin B_{12}
- **Protoctists** feed on the bacteria.
- All the microorganisms are rapidly growing and reproducing.
- Microorganisms eventually pass into the small intestine of the ruminant, die and are digested.
- The ruminant absorbs the resulting sugars, amino acids and vitamins.

Urea is often added to cattle feed. This is a source of ammonia for bacteria in their rumens and leads to more protein for the cattle to digest and amino acids to absorb.

✓ Quick check 4

Ruminants (such as cows) are in a sense 'farming' bacteria and protoctists, which they then feed on!

? Quick check questions

1. Suggest **two** reasons why two amylases are secreted by different parts of the gut.
2. Explain how the combined actions of several enzymes speeds up the digestion of protein in the gut.
3. Explain how the products of the digestion of fat are absorbed.
4. Antibiotics are drugs which are used to treat bacterial infections. Suggest why some antibiotics might harm cattle if given in tablet form.

B

HB

Control of digestive secretions

Digestion is under nervous and hormonal control; differences in the way these systems work are shown in the table.

Hormonal and nervous control of the digestive system

	Nervous control	Hormonal control
Nature of information	frequency of nerve impulses	concentration of hormone (a chemical)
How it is carried	along neurones directly to specific effectors	in the blood to all parts of the body
Number of targets	usually a specific part of the body – a muscle or gland	often targets at several sites – producing a number of responses
Speed of response	rapid – often within seconds	slower – minutes, hours or longer
Duration of response	usually short-lived	longer-lived, sometimes for hours, days or longer

Ⓢ The target cells of a hormone have specific protein receptors that the hormone binds to – because of their shape. The receptors may be on the cell membrane or inside the cell – depending on the hormone.

Nervous control of digestion

The autonomic nervous system (see page 78) is involved in these rapid responses.

Conditioned (learnt) reflexes

- Stimuli associated with food (sight, smell, taste) are detected by receptors which send nerve impulses to the brain.
- **Learning** leads to these stimuli being associated with food.
- The **sympathetic** part of the **autonomic nervous system** sends nerve impulses along the **vagus** nerve to the salivary glands, stomach and gallbladder.
- These produce saliva, gastric juice and bile in anticipation of arrival of food.

You make a conscious decision to eat, chew and swallow your food. After the food leaves your mouth, the rest of the process of digestion and absorption is under non-conscious control by reflexes, hormones and feedback mechanisms.

Simple reflexes

- When food is swallowed, it immediately stretches the stomach wall.
- Stretch receptors trigger a reflex causing release of hydrochloric acid into the stomach – to start digestion of polypeptides.
- Another reflex causes more muscular contractions (peristalsis) in the ileum – to move the contents along and make room for chyme from the stomach.
- When some chyme leaves the stomach, it stretches the wall of the small intestine (duodenum) and lowers the water potential of the gut contents.
- Receptors detect these stimuli and reflexes inhibit secretion of gastric juice, muscular contractions and release of chyme by the stomach – preventing chyme entering the small intestine until the first lot has been dealt with.
- Another reflex causes contraction of the gallbladder – releasing bile into the small intestine.

✓ Quick check 1, 2

Hormonal control of digestion

Hormones producing slower responses.

Gastrin is produced by certain cells of the stomach lining.

- If food contains short polypeptides or amino acids, gastrin is released into the blood and stimulates secretion of gastric juice with pepsinogen and HCl.
- **Positive feedback** – gastrin secretion increases the rate of digestion of protein, producing more short polypeptides and stimulating more gastrin secretion.
- **Negative feedback** – gastrin secretion causes the pH in the stomach to fall and at pH1 this causes all gastrin secretion to stop.

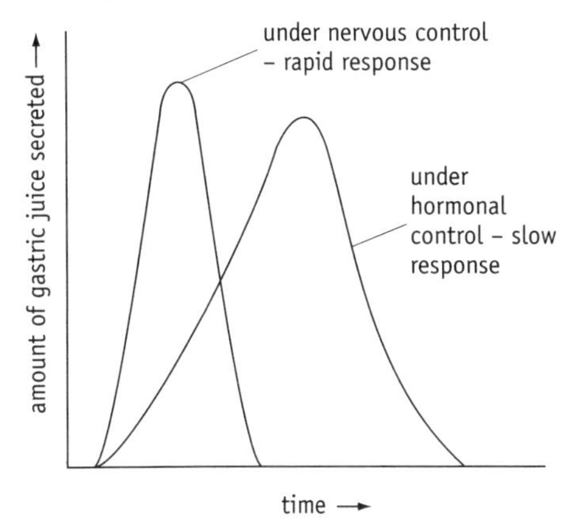

Comparison of speed of action of nervous and hormonal control on gastric juice secretion

Secretin is produced by certain cells of the small intestine (duodenum).

- **Acidic chyme** leaving the stomach lowers the pH in the small intestine (duodenum).
- This stimulus causes the secretion of secretin into the blood.
- Secretin causes release of pancreatic juice rich in **hydrogencarbonate ions**.
- This enters the small intestine (duodenum) and neutralises the chyme.
- **Negative feedback** – the rise in pH leads to no more secretin being released.

Cholecystokinin-pancreozymin is produced by certain cells of the small intestine.

- This is secreted into the blood when fat-rich chyme enters the small intestine.
- It causes release of pancreatic juice rich in digestive enzymes – lipase, amylase and trypsin.
- It also causes the gallbladder to contract – releasing bile that emulsifies fats and neutralises acidic chyme.
- **Negative feedback** – digestion of food leads to reduction in secretion of the hormone.

Secretin and cholecystokinin have different effects on cells in the pancreas, because they have different receptors on these cells which trigger different responses from the cells.

✓ Quick check 3

Quick check questions

1. Explain how the sight of food produces changes in gut secretions.
2. Explain how simple reflexes smooth the flow of chyme through the gut.
3. Explain the different effects of fat-rich and fat-poor meals on secretion of digestive juices in the small intestine.

B

HB

Histology of the ileum

Layers of the gut wall

These are similar throughout the gut but modified in different parts according to their role. Absorption of digested food products occurs to a great extent in the **ileum**; part of the **small intestine**.

The ileum is **adapted** for **absorption** by having:

- a **large surface area** due to its long length, folds in the wall and **villi**,
- villi covered by a **single layer** of epithelial cells – providing a **short diffusion pathway**,
- **blood capillaries** and **lacteals** in villi – carrying away absorbed substances to maintain **high diffusion gradients.**

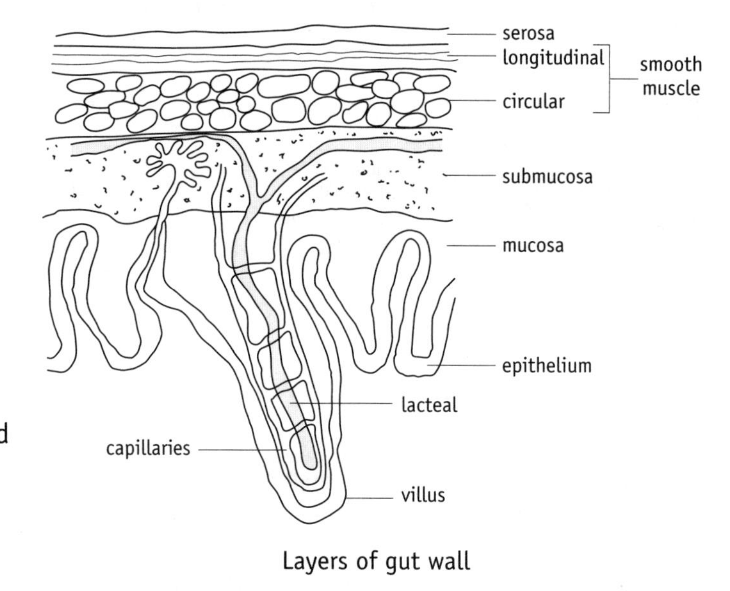

Layers of gut wall

Its **epithelial cells** have:

- the cell membrane facing the gut lumen folded into **microvilli**, giving an increased surface area for diffusion, facilitated diffusion and active uptake,
- **carrier proteins** in the cell membrane – increasing permeability,
- many **mitochondria** to supply **ATP** for **active uptake**.

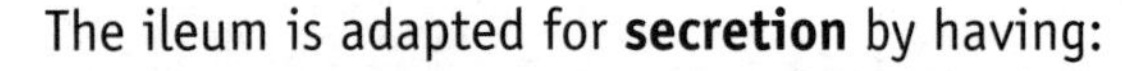

The ileum is adapted for **secretion** by having:

- intestinal crypts (crypts of Lieberkühn), which produce intestinal juice – containing water, digestive enzymes and mucus,
- paneth cells formed in the crypts produce digestive enzymes,
- goblet cells that secrete mucus.

Quick check 1, 2

Quick check questions

1 Explain how the ileum is adapted for the absorption of the products of digestion.

2 Explain how epithelial cells of the ileum are adapted for their roles in digestion.

Secretory cells in the gut epithelium have lots of mitochondria – to supply energy for synthesis of enzymes or mucus. They have lots of rough endoplasmic reticulum to synthesise enzymes.
They have Golgi bodies – to package proteins or make mucus.

B

Metamorphosis and insect diet

Metamorphosis in insects is a **change of form** or structure during its life cycle. These forms use different food sources, occupy different habitats, and thus reduce intraspecific competition.

Life cycle of a butterfly

Butterflies belong to the order **Lepidoptera** (butterflies and moths). They undergo complete metamorphosis; the **larva** (caterpillar), **pupa** (chrysalis) and **adult** (imago) have very different body forms.

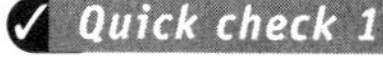

The larva

- Eggs are laid on or near a host plant – the larva does not need to move much to find food.
- The **larva is the main feeding and growth stage.**
- It feeds on leaves using its chewing mouthparts and **grows rapidly.**
- Growth needs a **lot of energy from respiration** and **protein synthesis.**
- The larva produces enzymes to digest starch, sucrose and lipids in its food, to give simple sugars and fatty acids for respiration.
- It produces enzymes to digest protein, to give amino acids for protein synthesis.
- When growth is complete, the larva changes into a pupa which does not feed.
- Inside the **pupa, larval tissues are broken down, releasing biological molecules that are used to make the adult butterfly.**

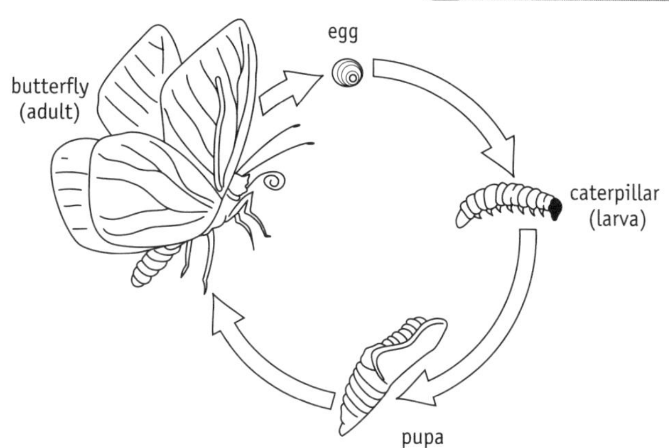

Life cycle of a butterfly

Ⓢ Make sure that you revise the biochemistry of respiration, including RQ values. Also make sure that you know which small biological molecules are used to make large biological molecules.

The adult

- The **adult does not grow, its function is to find a mate to reproduce.**
- This means that it only needs food that provides energy for flying.
- It feeds on liquid nectar using a long, sucking proboscis.
- Nectar is produced by flowers and contains a lot of dissolved sucrose.
- The butterfly only has the enzyme to digest sucrose.

Enzymes produced in the guts of the larval and adult stag

Stage of life cycle	Diet	Protease	Lipase	Amylase	Sucrase	Maltase
Larva (catterpillar)	leaves	✓	✓	✓	✓	✓
Adult (butterfly)	nectar	✗	✗	✗	✓	✗

✓ Quick check 2

? Quick check questions

1. Name the main stage in the life cycle of an insect which undergoes complete metamorphosis.
2. Explain why the larva of the butterfly produces a large range of digestive enzymes but the adult does not.

Ⓢ Different genes must be switched on in the larva and the adult, so that different enzymes are produced.

B

HB

Neurones and action potentials

Neurones

Neurones generate **nerve impulses**, carrying information along axons.

Very thin extensions carry nerve impulses towards and away from the cell body.

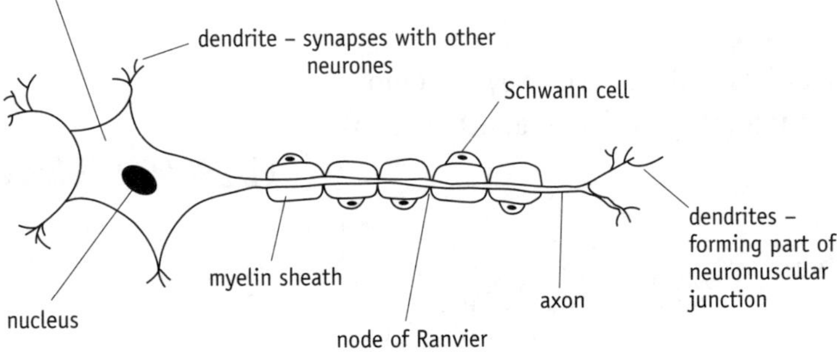

Myelinated motor neurone

Ⓢ You have to know the various ways in which substances enter and leave cells across cell membranes. You also need to know how the structure of the membrane is related to its functions.

Motor neurones carry nerve impulses to muscles fibres at neuromuscular junctions, leading to contraction.

- The axon of the motor neurone in mammals is myelinated.
- The myelin sheath of fatty (cell membrane) material is made by Schwann cells; each making a few millimetres of the sheath.

✓ *Quick check 1*

Resting potential

There is a potential difference across the cell membrane of a resting neurone of about **-70 mV** (millivolts) – the membrane is **polarised.**

- The inside is negative compared to the outside.
- This is a due to a higher concentration of **sodium ions** outside the cell.
- The cell membrane is usually impermeable to sodium ions.
- Some sodium ions do diffuse in, but are **actively transported out** by a transmembrane **carrier protein** called a **cation pump (sodium pump).**
- **Potassium** ions are actively transported into the cell.
- Each type of pump is a protein with a tertiary structure which allows it to recognise and bind to a specific cation.

Use the **correct terminology** – many candidates do not use words like, depolarized, active transport, sodium channels, tertiary structure, refractory period, resting potential and nerve impulse. Other, vaguer words will **not** get credit.

Action potential

An action potential is a **travelling depolarisation** of the cell membrane; the resting potential disappears for a few milliseconds. An action potential starts at the cell body and travels to a synapse.

B

HB

- A stimulus causes transmembrane protein channels to open.
- Sodium ions diffuse in along their concentration gradient, causing a rapid, local depolarisation of the membrane.
- The inrush of sodium ions makes potassium channels open, allowing potassium ions to diffuse out and start to restore the resting potential.
- Sodium channels then close and the sodium pump starts to pump sodium ions out of the cell – to restore the resting potential.
- The local depolarisation opens sodium channels in the next section of the membrane – so the action potential travels.

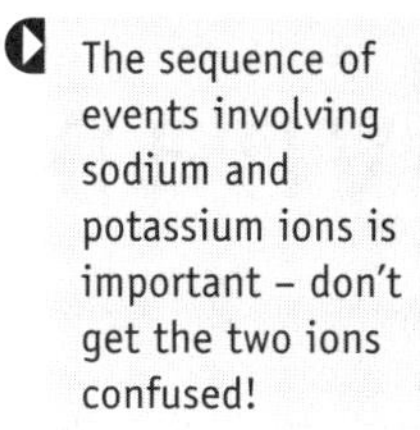

An action potential:

- only happens if a stimulus reaches a **threshold** value,
- it is an **'all or nothing event'** – once it starts, it travels to a synapse,
- is always the **same size** – one travelling action potential is a **nerve impulse**.
- The **frequency of nerve impulses carries information**; a strong stimulus produces a high frequency of nerve impulses.
- After depolarisation, there is a time when no new action potential can start – the **refractory period** – this produces **discrete/separate nerve impulses**.

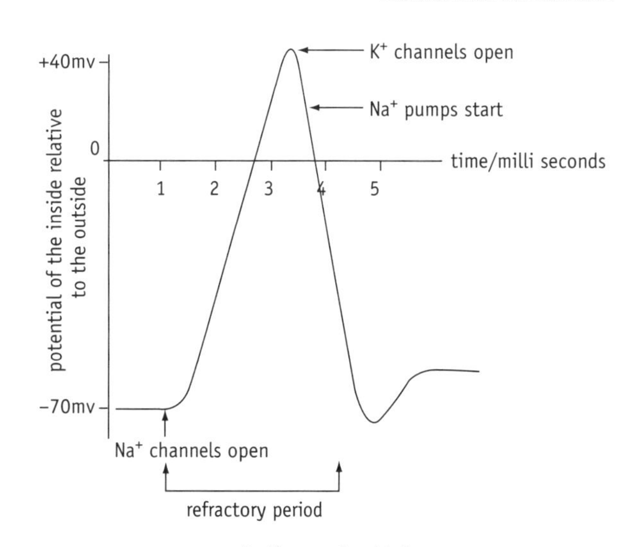

Action potential

Speed of conduction

Myelination, **diameter** of axon and **temperature** affect speed of conduction.

- **Myelinated** axons are found in humans (and other vertebrates), invertebrates have non-myelinated axons.
- In myelinated axons, nerve impulses 'jump' **instantly** from node to node, where the axon membrane is exposed.
- This **saltatory conduction** makes speed of travel of the nerve impulses faster than in non-myelinated axons.
- The greater the diameter of the axon, the greater the speed of conductance.
- The higher the temperature, the greater the speed of conductance.
- This is because action potentials depend upon diffusion of ions, opening of ion channels and active transport – all of which are temperature sensitive.

Speed of conductance shows a Q_{10} of about 1.8 – the rate increases by about 1.8 for every 10°C rise in temperature. Mammals maintain a constant body temperature, so their speed of conduction will stay the same.

Quick check 4

Quick check questions

1. Explain how the structure of a motor neurone is adapted to its function.
2. Explain how an action potential travels along an axon.
3. Use the information in the graph above to calculate the maximum number of nerve impulses per second that the axon could carry.
4. We do not react to very small stimuli and there is a limit to our ability to detect the size of very large stimuli. Suggest how these characteristics are linked to the formation and passage of action potentials.

B

HB

Synaptic transmission

Synapses are where neurons communicate with each other, or with an effector; a muscle or gland.

- There is a **synaptic cleft** (about 20 nm) between cell membranes of neurones.
- Information is carried across the cleft by chemical **neurotransmitters.**

Acetylcholine is the neurotransmitter in:

- the **parasympathetic** part of the **autonomic nervous system,**
- **neuromuscular junctions** in the voluntary nervous system,
- **cholinergic** synapses.

Noradrenaline is the neurotransmitter in;

- most of the **sympathetic** autonomic nervous system.
- **adrenergic** synapses.

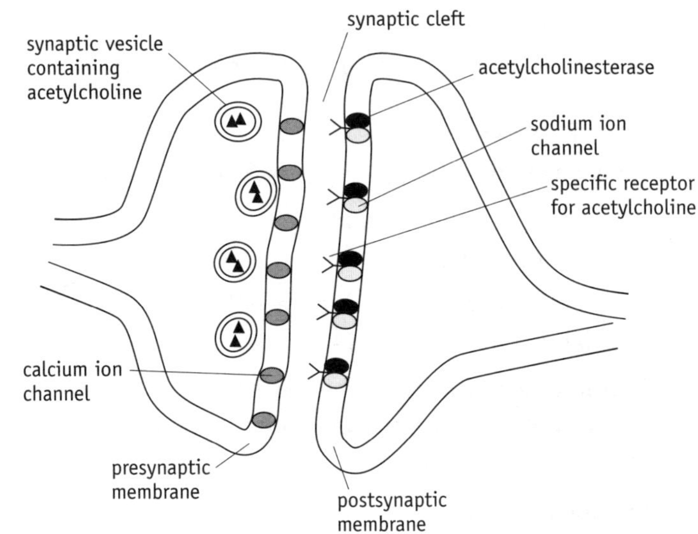

A synapse

Adrenaline is a hormone produced by the adrenal glands at times of stress. It can produce the same effects in the sympathetic nervous system as noradrenaline.

Excitatory cholinergic synapses and unidirectionality

- An action potential reaches the **presynaptic membrane** of a synapse.
- Depolarisation makes **calcium channels** open and **calcium** ions diffuse **in.**
- Calcium ions cause **synaptic vesicles** containing acetylcholine to **fuse** with the membrane, releasing acetylcholine into the synaptic cleft.
- Acetylcholine **diffuses** very quickly across the synaptic cleft and binds (reversibly) to **specific receptors** on the **postsynaptic membrane.**
- The receptors are proteins whose **tertiary structure** only fits acetycholine.
- Binding opens **sodium channels**, sodium ions diffuse in, causing **depolarisation** of the postsynaptic membrane (excitatory postsynaptic potential).
- If enough depolarisations are produced frequently enough, an action potential occurs in the postsynaptic neurone.
- **Synaptic transmission is unidirectional** – information is carried in one direction.
- There are **no receptors on the presynaptic membrane**, so acetylcholine can not cause depolarisations there.
- Acetylcholine is rapidly broken down by an enzyme, **acetylcholinesterase.**
- Otherwise, it would keep attaching to receptors causing unwanted depolarisations.

Many candidates do not use the correct terms when writing about synapses and miss out important events; such as fusion of vesicles with the presynaptic membrane.

A series of action potentials arriving at a synapse is linked to bursts of release of transmitter substance into the synaptic cleft and these bursts are linked to the number and frequency of action potentials in the postsynaptic neuron.

✓ Quick check 1, 2

B

HB

Inhibition

- In **excitatory** synapses, acetylcholine binds to a type of receptor that makes **sodium channels open**, causing **depolarisation** of the postsynaptic membrane.
- In **inhibitory** synapses, acetylcholine binds to a different type of receptor that makes **potassium channels open**.
- Potassium ions diffuse out, causing **hyperpolarisation** of the postsynaptic membrane – making it more difficult to **depolarise** and **preventing an action potential**.

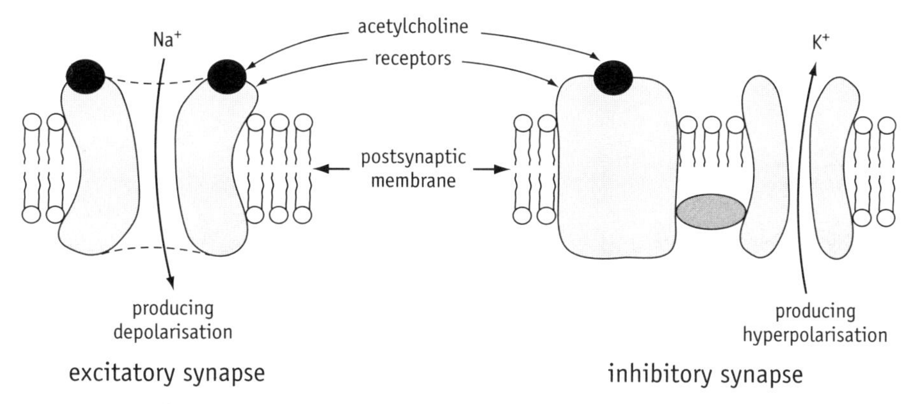

excitatory synapse inhibitory synapse

Nicotine acts in the central nervous system by binding to acetylcholine receptors.

Summation can involve excitatory and inhibitory synapses. The inhibitory synapses can subtract from the additive effects of excitatory synapses.

Ⓢ You do not need to know specific examples of drugs, so questions will contain the information needed to answer them.

Ⓢ Look very carefully at any diagrams in questions about drugs and synapses – look for corresponding shapes.

Summation

Most neurones synapse with many other neurones. A single postsynaptic depolarisation at one synapse is unlikely to produce an action potential.

- Several depolarisations occuring closely together have an additive effect, reach a threshold, and lead to formation of an action potential.
- **Spatial summation** additive effect of depolarisations from two or more synapses.
- **Temporal summation** additive effect of depolarisations produced rapidly one after the other at the same synapse.

✓ Quick check 3

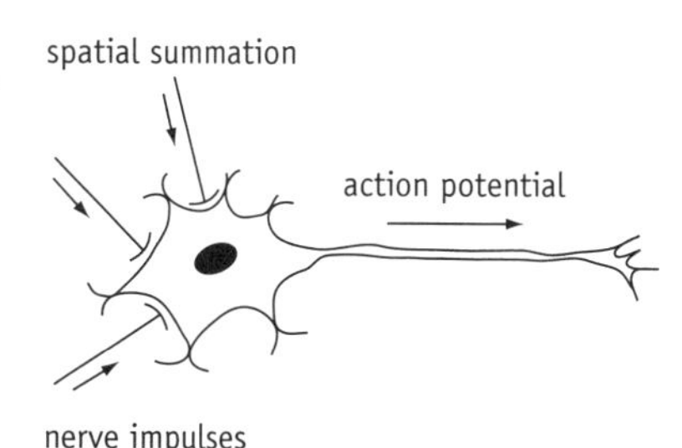

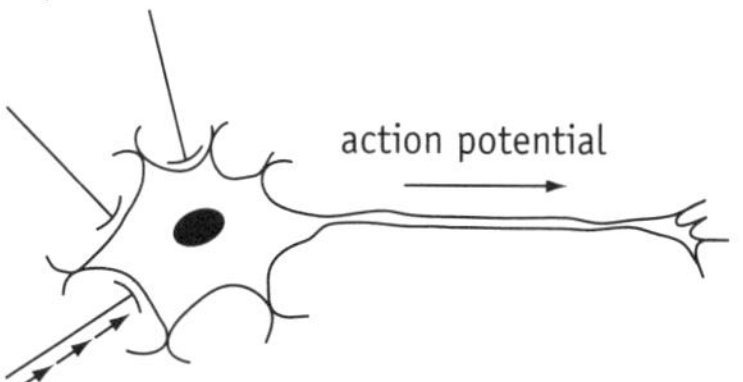

Summation

✓ Quick check 4

Agonistic and antagonistic effects of chemicals

- **Agonists** are substances with shapes that allow them to bind to the same receptors as neurotransmitters and mimic their action – producing the response.
- **Antagonists** block the normal action of neurotransmitters – preventing/reducing a response.
- Some drug molecules **compete** to bind to neurotransmitter receptors.
- This might prevent depolarisation of the membrane, or make depolarisations happen all the time.
- Some drugs inhibit acetylcholinesterase, so acetylcholine remains in the synapse, producing continuous depolarisations.

? Quick check questions

1. Explain how information is carried across a synapse.
2. Explain why acetylcholinesterase is necessary in a synapse.
3. Neurones often have inhibitory and excitatory synapses with other neurones. Suggest why the result is that they sometimes produce action potentials and sometimes they don't.
4. A snake venom paralyses prey. The shape of the venom molecule is shown below. Use the information in the diagram of a synapse to suggest how the venom has its effect.

B

HB

Receptors

Receptors are specialised cells which detect a specific **stimulus**; a change in the internal or external environment of an organism. A stimulus above threshold value produces nerve impulses. Receptors are transducers, changing the energy of a stimulus into a form the organism can understand. Receptors may be grouped together in a sense organ such as the eye.

A stimulus involves a change in energy, e.g. light, touch or sound. The nature of the nerve impulse is always the same.

Pacinian corpuscles

These are mechanoreceptors which;

- are found deep in the dermis of the skin, some joints and tendons,
- are sensitive to pressure and vibrations,
- consist of the ending of a single neurone surrounded by many **lamellae**.

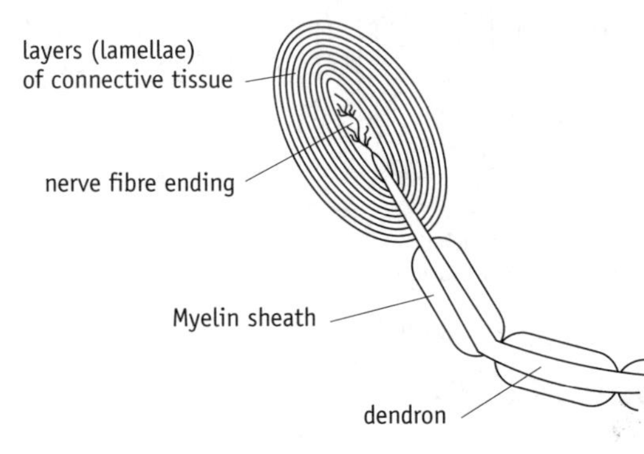

A Pacinian corpuscle

Transduction in a Pacinian corpuscle

- Pressure on the skin changes the shape of the Pacinian corpuscle.
- This changes the shape of **pressure sensitive sodium channels** in the membrane, making them open.
- Sodium ions diffuse in through the channels leading to a depolarisation called a **generator potential.**
- The greater the pressure, the more sodium channels open and the larger the generator potential.
- If a threshold value is reached, an action potential occurs and a nerve impulses travels along the sensory neurone.
- The frequency of impulses is related to the intensity of the stimulus.

Quick check 1

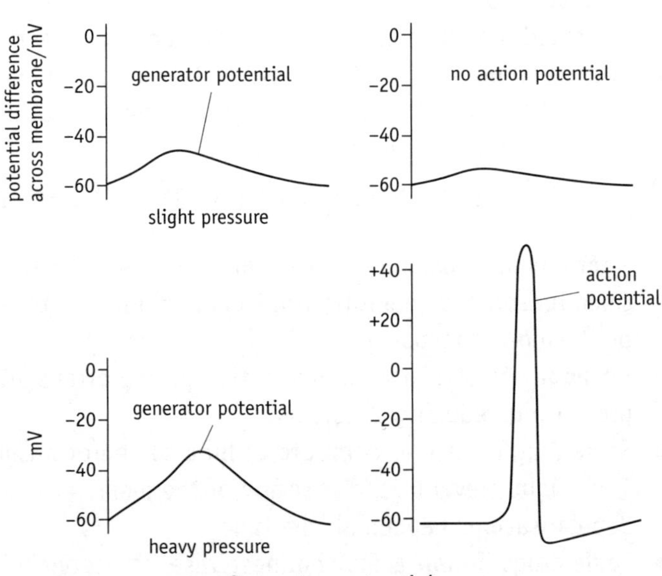

Generator potentials

Quick check 2

Quick check questions

1 Where are Pancinian corpuscles found in the body?

2 Explain why the body may not respond to a slight pressure on the skin.

The tertiary structure, the shape, of proteins is always linked to their function. You need to know how changes in protein shape is related to substances crossing membranes – facilitated diffusion and active transport.

B

HB

The retina

The eye is a **sense organ**. It has **receptor cells** called **rods** and **cones** in the retina. Light reflected from an object reaches the eye and is focused on the retina.

Remember! The cornea carries out most of the focusing (refraction) of light onto the retina and the lens carries out fine focusing.

Focusing

- A clear image of an object is focused on the **fovea** of the retina.
- **Most focusing** (refraction, i.e. bending of light) is done by the curved surface of the transparent **cornea.**
- The shape of the **lens** can be altered to carry out **fine focusing** – **accommodation.**
- Looking at distant objects, lens kept thin by **tension** in **suspensory ligaments.**
- To look at near objects, the circular **ciliary muscles** contract, taking tension out of the ligaments – the natural elasticity of the lens makes it fatter (more biconvex).

Quick check 1

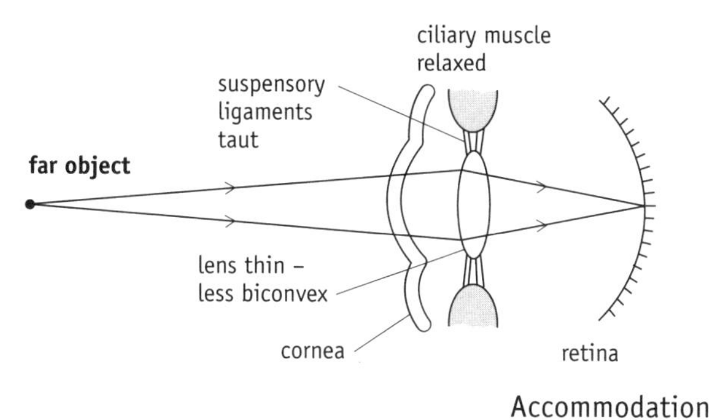

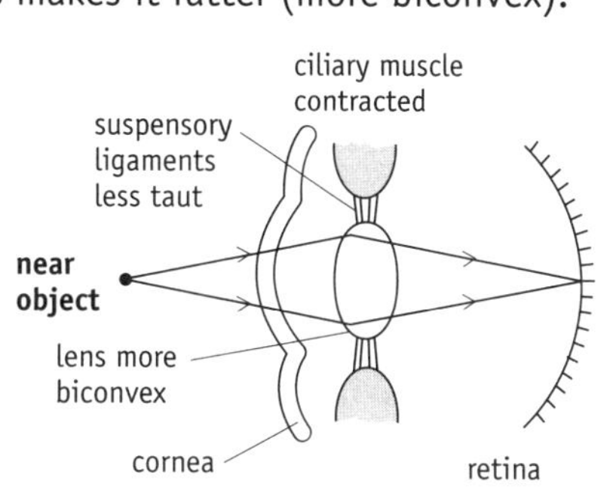

Accommodation

Don't learn a lot of unnecessary biochemical detail about the production and turnover of rhodopsin!

Rods and cones

Rods and cones have different structures and light-sensitive pigments.

- Rods contain the light-sensitive pigment **rhodopsin** as part of the structure of membranes of vesicles in the outer segment of the rod cell.
- Light 'bleaches' rhodopsin, making it break down to retinene and scotopsin.
- This alters the permeability of the membrane to Na^+ ions, leading to formation of nerve impulses that pass along the optic nerve to visual centres in the brain.
- In the absence of light, rhodopsin is regenerated from retinene and scotopsin.

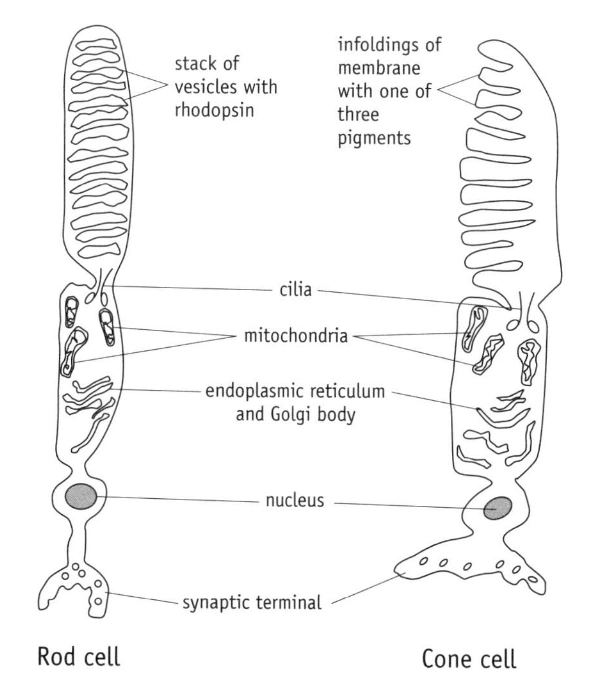

Quick check questions

1. Explain what happens to maintain a clear image on you retina when you change from looking at a distant object to reading a book.
2. Explain how the stimulus of light falling on a rod cell is converted into information the body can use.

B

HB

Monochromatic and colour vision

Rods give **monochromatic** (black and white) vision without a great deal of detail. They are found outside the fovea, have high sensitivity to light and work in low light intensities.

Cones are used for **detailed colour** vision. They are found in the fovea of the retina, have a relatively low sensitivity to light – only work in high light intensities.

Make sure you write about cone cells with a pigment showing maximum absorption in the red part of the spectrum - **not** 'red cones'!

Colour vision

The **trichromatic theory** of colour vision is based upon **three types of cone cell** found in the retina.

- They are blue-sensitive, green-sensitive, or red-sensitive, depending on the wavelength at which they show maximum absorption.
- Each type contains a **different pigment.**
- Other colours cause reactions in combinations of cones. So, for example, brown light stimulates green-absorbing and red-absorbing cones.
- Red-green colour blindness is caused by a lack of either green-absorbing or red-absorbing cones.

Quick check 1

Colour blindness is due to an allele of a gene which produces a different pigment, compared to the normal pigment. This allele is the result of a mutation.

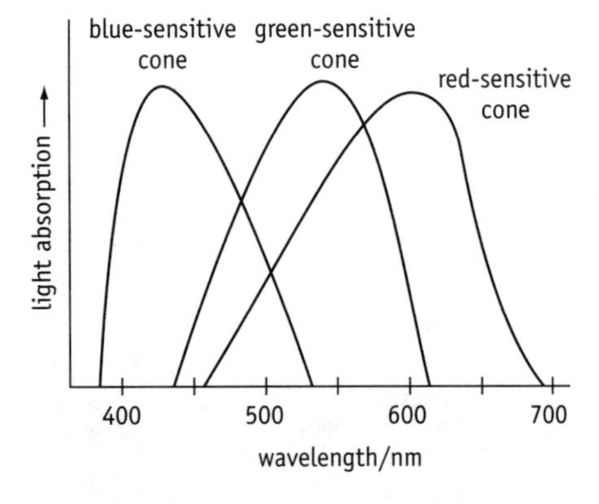

Absorption spectra of cone cells

Sensitivity and visual acuity

When we look directly at something, its image falls on the fovea and we see it in colour and sharp detail. Objects in the periphery (edge) of our field of view are not seen in colour, or detail. The fovea has a very high density of cones.

- Each **cone** cell has a synapse with one bipolar cell and one ganglion cell.
- So each cone sends nerve impulses to the brain about its own small area of the retina – giving high **visual acuity**.

Rod cells provide peripheral black and white vision.

- **Rod** cells are connected in groups to one bipolar cell and one ganglion cell.
- Groups of rods send information to the brain about a relatively large area of the retina (compared with individual cone cells).
- If light from an object falls anywhere inside the area of a group of rod cells, the brain is aware of this but the view of the object will not be very sharp – the **visual acuity is low.**

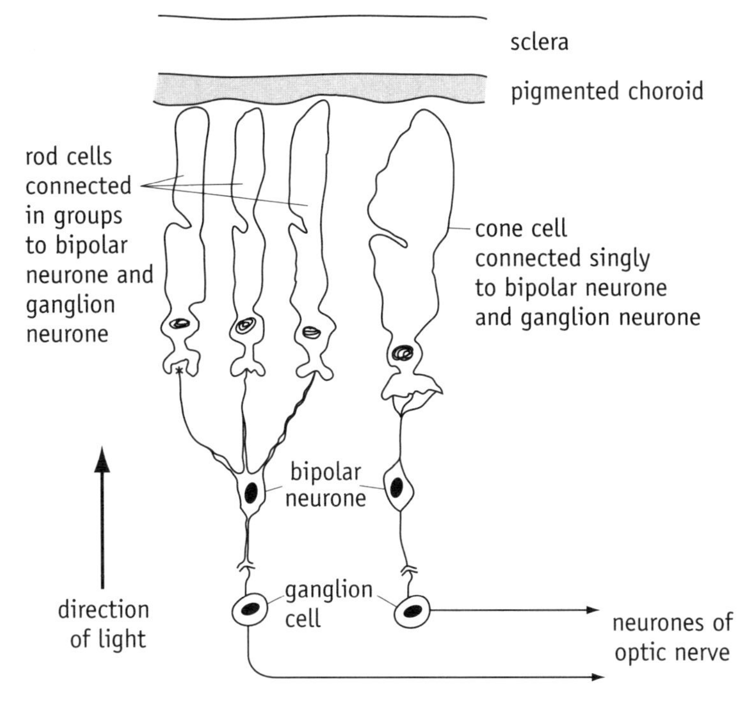

Structure of the retina

✓ Quick check 2,3

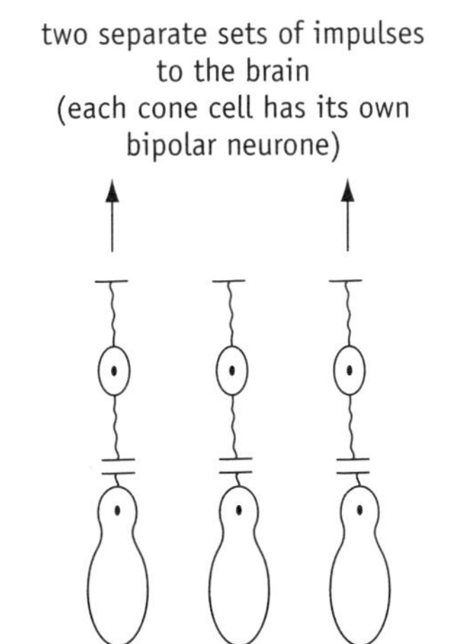

cones – high visual acuity

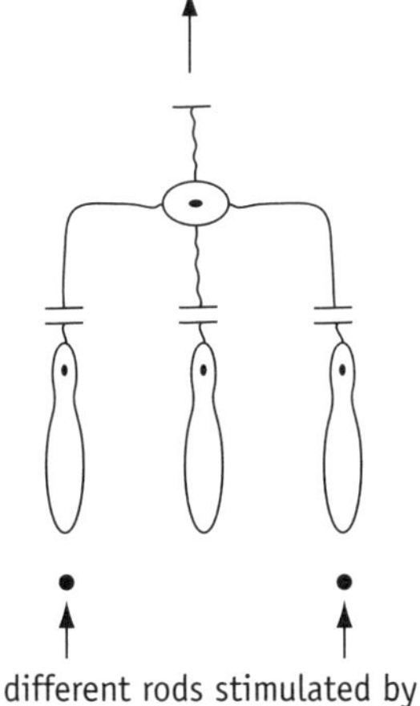

rods – low visual acuity

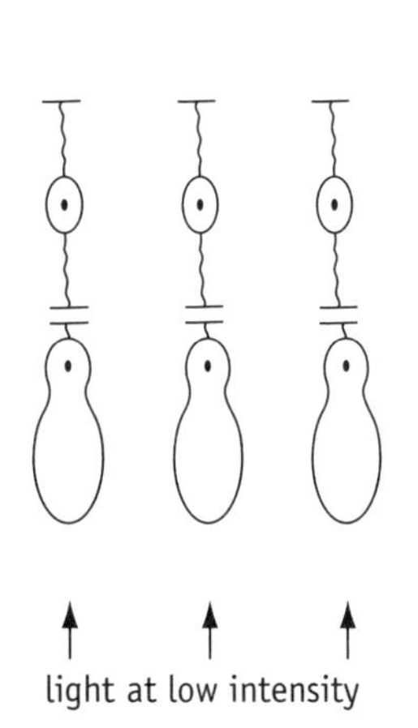

cones – low sensitivity

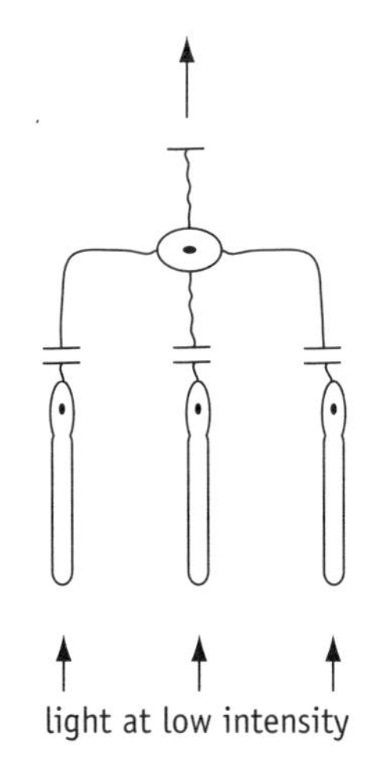

rods – high sensitivity

Sensitivity and visual acuity

Quick check questions

1. Some people who are red-green colour blind have three types of cone cell but the pigments of two of the types have very similar absorption spectra. Sketch a graph similar to the one on page 76, to show the absorption spectra of the cones of such a person.
2. At night, we often see moving things out of the corner of our eye, but cannot see them when we look directly at them. Suggest why this happens.
3. Explain why we see so much detail when we look at coloured illustrations.

Make sure you understand the links between the ways rods and cones are connected to ganglion cells and visual acuity.

B

HB

Autonomic nervous system

The autonomic nervous system controls functions that are **involuntary/ non-conscious**, including **homeostatic** systems. Overall control is by the **hypothalamus** and **medulla** of the brain. Local control is through reflexes controlled by ganglia (clusters of neurones) outside the central nervous system. Autonomic control involves a stimulus, receptor, coordinator and effector.

Spinal reflex

A reflex is an **automatic** response. Its **adaptive value** is that it **protects** an organism from a harmful stimulus. The **hand withdrawal reflex** happens if you put your finger onto something which is too hot. This involves **three neurones.**

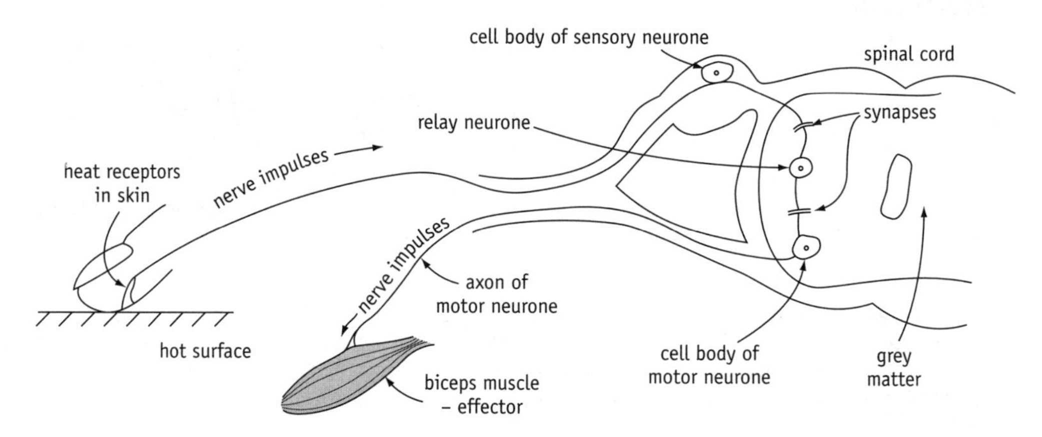

A simple reflex

- Heat **receptors** detect a harmful rise in skin temperature, the **stimulus.**
- Nerve impulses travel along a **sensory neurone** into the spinal cord (the **coordinator**), to a synapse with a **relay/association neurone.**
- The relay neuron synapses with a **motor neurone**, which sends nerve impulses to the **effector**, the biceps muscle.
- This contracts pulling the hand away from the heat – the **response.**

Always use the terms used here. For example, nerve impulses are not 'messages'!

Quick check 1

Sympathetic and parasympathetic

- The **sympathetic** and **parasympathetic components** of the autonomic nervous system have **opposing effects** on the body.
- The sympathetic prepares the body for 'flight or fight', e.g. by increasing heart rate and breathing rate, increasing blood supply to muscles and reducing blood supply to the skin and gut.
- **Noradrenaline** is the neurotransmitter in sympathetic synapses.
- The parasympathetic system uses **acetylcholine** and reverses the effects listed above.

Make sure that you don't confuse the actions of the sympathetic and parasympathetic systems, or their neurotransmitters.

B

HB

Examples of autonomic control

Target	Sympathetic effect	Parasympathetic effect
pupil of eye	dilates	constricts
bronchi	dilation	constriction
heart	increase in rate	slowing of rate
blood vessels	vasoconstriction	vasodilation
blood supply to gut	reduced	increased

Control of heart rate

The heart beat is initiated by the sinoatrial node (SAN). The rate at which the SAN operates can be changed to meet the demands of the body.

- **Sympathetic** and **parasympathetic** (vagus) neurones connect to the SAN.
- They are always active and alter the rate of depolarisation of the SAN.
- **Noradrenaline** released by sympathetic neurones speeds up the SAN.
- **Acetylcholine** from parasympathetic neurones slows the SAN.

During exercise **heart rate increases** because of:

- **fewer nerve impulses** in **parasympathetic** (vagus) neurones,
- **reducing inhibition** of the rate of depolarisation of the SAN,
- and **more nerve impulses** in **sympathetic** neurones,
- which stimulate the SAN.
- Heart rate is controlled by a **cardiac centre** in the **medulla** of the brain: divided into a cardioaccelerator centre and a cardioinhibitor centre.
- Contracting muscles pressing on veins forces blood towards the heart, causing greater filling of the ventricles which makes the heart beat faster and stronger.
- If blood pressure rises too far above normal, **pressure receptors** in the aorta and carotid sinus send nerve impulses to the cardioinhibitor centre.

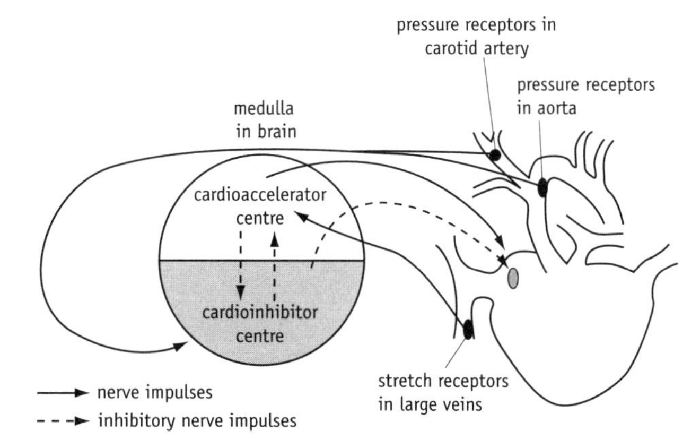

Changes in heat rate with increase in muscular activity

- This centre sends inhibitory nerve impulses to to the cardioaccelerator centre and the SAN – preventing the heart beating too fast.

✓ Quick check 2

Adrenaline is a hormone secreted by the adrenal glands (on top of the kidney). This acts on sympathetic synapses in the same way as noradrenaline – producing very similar effects.

Sympathetic neurones to the atria and ventricles increase the strength of contraction and very slightly reduce the time in systole.

Ⓢ The heart is myogenic, heart beats are intitiated by electrical activity in the SAN, followed by the atrioventricular node and atrioventricular bundle.

Don't confuse how heart beat is regulated with the control of breathing!

✓ Quick check 3

Quick check questions

1. Explain why we pull a finger away from a hot surface before we ever realise it is hot.
2. Suggest how **two** of the examples of sympathetic effects in the table on autonomic control help in a 'fight or flight' reaction to a frightening stimulus.
3. Explain why heart rate increases during exercise.

B

HB

Behaviour

Taxes and kineses

Behaviours have evolved that improve the chances of survival of an organism; often by making it find, or stay in, a favourable environment. Taxes (singular: taxis) and kineses (singular: kinesis) are examples of **innate behaviour** that is **genetically determined**, not learned. Members of a species which **inherit** the same allele(s) for a behaviour produce the same response to a stimulus.

Ⓢ There is variation in behaviour in a population; due to genetic and environmental factors. Selection will take place - directional, stabilising or disruptive. The organisms with the most advantageous behaviour are more likely to survive, reproduce and pass on their alleles.

Taxis

- Involves the movement of the **whole organism**.
- Movement may be towards (positive) or away (negative) from a **directional stimulus**.
- The direction of the response is related to the direction of the stimulus.
- Taxes are classified according to the stimulus involved.

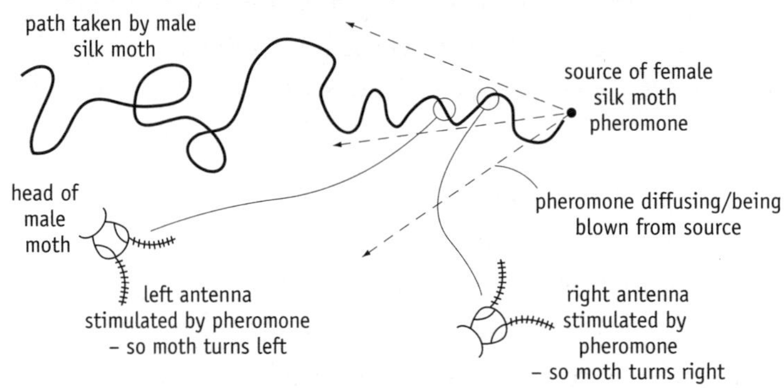

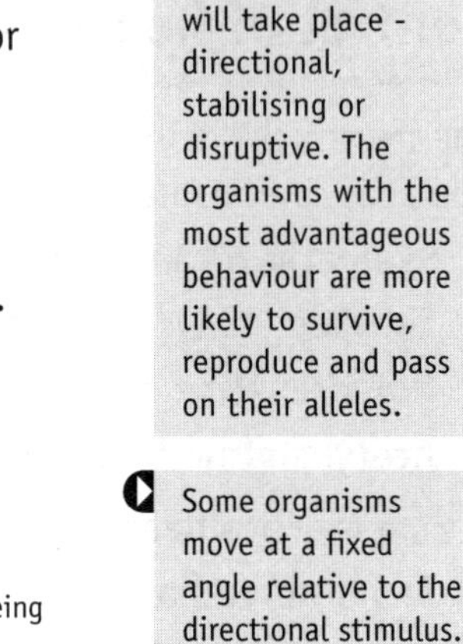

Positive chemotaxis

Some organisms move at a fixed angle relative to the directional stimulus.

Taxes are usually associated with invertebrates and their behaviour patterns

Type of taxis	Type of stimulus	Example
phototaxis	light	maggots moving away from light (negative photaxis)
chemotaxis	chemical e.g. pheromones	male moths flying towards pheromones from females (positive chemotaxis)
aerotaxis	oxygen (air)	motile aerobic bacteria moving towards oxygen (positive aerotaxis)

Kinesis

- Involves a change in the **rate of movement** or activity of an organism.
- Rate of movement, e.g. turning, is related to intensity of the stimulus.
- Movement of the organism is random and not related to the direction of the stimulus.
- Choice chambers can be used to investigate kinesis in some species, e.g. woodlice.

Woodlice show rapid movements in dry conditions but are less active in humid conditions. This aids survival - the woodlice stay in areas where dehydration is less likely.

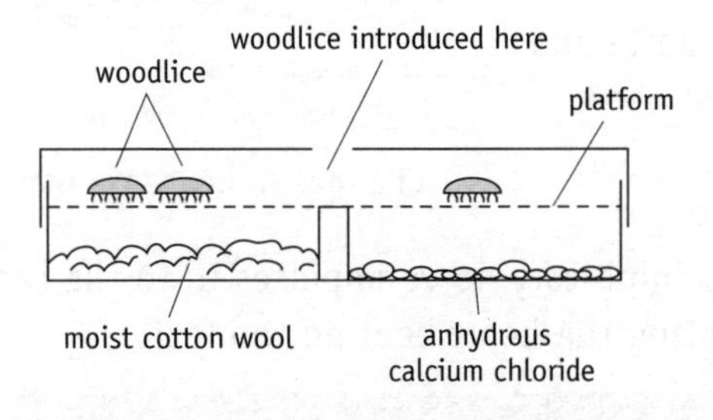

Use of choice chambers

Quick check questions

1. Earthworms move away from light. Name this type of behaviour and suggest how earthworms benefit from this response.
2. Suggest two conditions which should be kept constant when investigating the effect of humidity on the behaviour of woodlice.

B

Module 6: end-of-module questions

1 a The diagram shows a cross-section of a root.

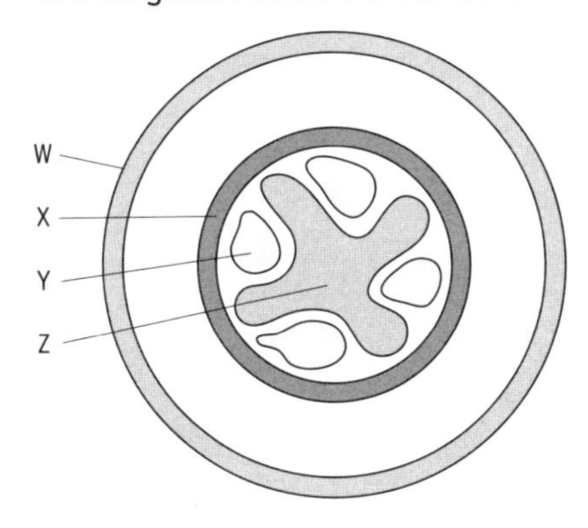

Identify **W, X, Y** and **Z.** [4]

b Explain how water moves into the xylem in the root from the soil. [6]

2 a Explain how a water molecule gets from the xylem in a root to the air outside stomata. [8]

b Describe and explain how **three** environmental factors affect transpiration. [6]

c Explain two ways in which the leaf of a xerophyte may be adapted to reduce transpiration. [4]

3 The diagram shows a liver cell, its receptors for insulin and glucagon and an outline of the metabolic pathways involving glycogen.

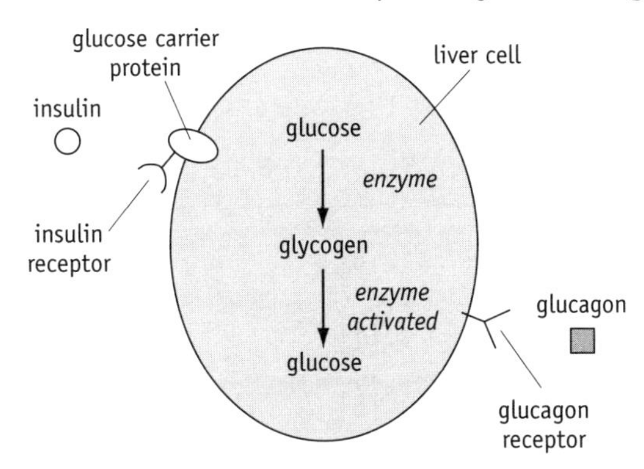

a Use information in the diagram and your own knowledge to explain:

i how the release of insulin by the pancreas leads to a reduction in glucose in the blood. [3]

ii how glucagon release leads to the release of glucose into the blood. [2]

B

b Some people inherit an allele which produces a faulty enzyme in the pathway between glycogen and glucose. Suggest what symptoms they would show. [2]

4 There is a plant found in Australia which produces different types of leaf. The first type is produced in the spring, which is relatively cool and wet. The second type is produced in the summer, which is very hot and dry. The chart shows the characteristics of the two types of leaf.

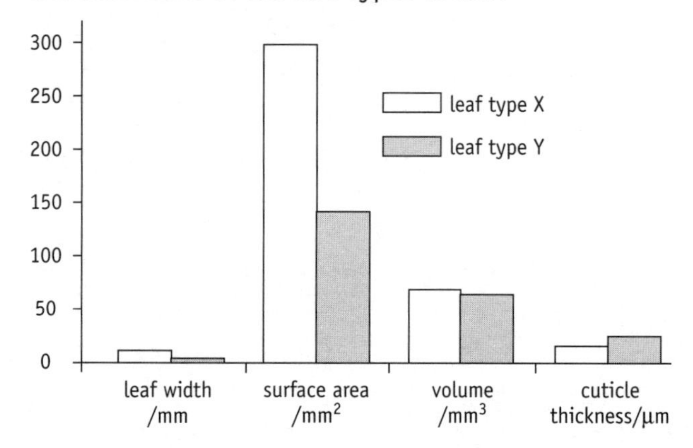

a Assume that concentrations of carbon dioxide and oxygen in the atmosphere are the same in spring and summer. Use the data to explain which leaf type is better adapted for gaseous exchange. [4]

b Suggest which type of leaf is produced in the summer. [3]

5 The flow-chart shows some of the pathways involved in the formation of nitrogenous waste and the form in which it is excreted by different organisms.

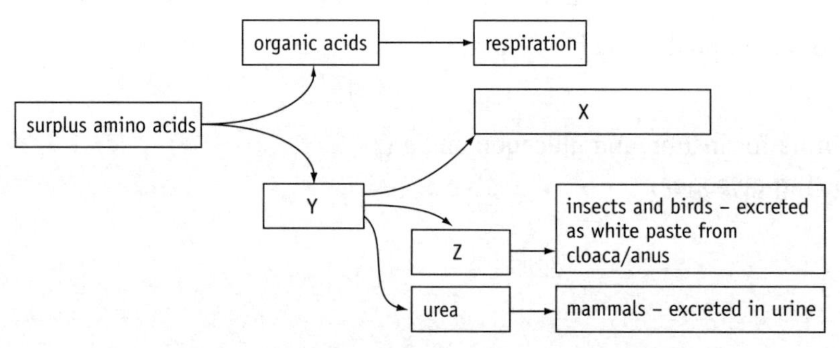

a Name compounds **Y** and **Z**. [2]

b Name one type of organism that could represent **X**. [1]

c Explain how gomerular filtrate is formed in the renal capsule. [5]

d The table shows the water budget of a desert mammal.

Water budget of desert mammal for one month	
Water gains	**Water losses**
Oxidation water from metabolism – 90%	Urine – X%
From carbohydrate from seeds – 10%	Faeces – 0.043%
	Evaporation from lungs and skin – 73.2%
Total water gain $60cm^3$	Total water loss $60cm^3$

i Explain where the oxidation water comes from. [2]

ii Calculate how many cm^3 of water were lost as urine. [2]

6 a Explain how and where proteins are digested in the human gut. [5]

b Describe how the release of pancreatic secretions is controlled. [4]

7 The graph shows the changes in potential across part of the membrane of the axon of a motor neurone during an action potential.

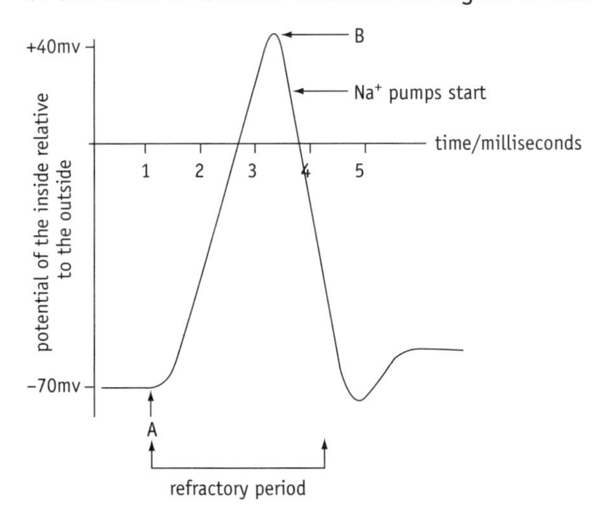

a Explain what happens at **A.** [3]

b Explain what happens at **B.** [3]

c Use the graph to calculate the maximum frequency of nerve impulses passing along this axon. [2]

8 a Other than the control of heart rate, explain **two** differences between the sympathetic and parasympathetic nervous systems in terms of their effects on human physiology. [4]

b Explain how heart rate is increased during exercise. [4]

HB

Module 7: The Human Lifespan

This module is broken down 12 topics: Sexual reproduction, The developing fetus and maternal system, Growth and development, The digestive system*, Dietary requirements, Transport of respiratory gases*, Neurones, action potentials and synapses*, Receptors*, Muscles, Reflexes and the autonomic nervous system*, Homeostasis* and Senescence.

Some sections* are common to Module 6 (Biology) and are covered in that module – where they are indicated by HB

Sexual reproduction

Gametes are produced in the ovaries and testes in processes involving mitosis, meiosis, growth and maturation. Gametes fuse at fertilisation to form a zygote.

The developing fetus and maternal system

A zygote develops into a blastocyst embryo which implants in the uterus. Fetal blood circulation is adapted to having the placenta as the exchange surface for all essentials. Pregnancy, birth and lactation are controlled by hormones, which affect maternal physiology.

Growth and development

Absolute growth and growth rate can be measured. Different parts of the body grow at different times and rates. Important hormonal changes and growth occur at puberty.

Digestion and absorption B 

Dietary requirements

Principle nutrients include carbohydrates, lipids, proteins, vitamins and mineral ions. The basal metabolic rate reflects energy needs of the body. This varies depending upon age, sex and physical demands placed on the body.

Transport of respiratory gases  

Neurones, action potentials and synapses B HB

Receptors  

Muscles

A muscle consists of muscle fibres. These contain myofibrils. Contraction is due to interactions between actin and myosin which require ATP and are described by the sliding-filament hypothesis.

Reflexes and the autonomic nervous system B HB

Homeostasis B 

Senescence

Ageing produces a decline of physiological functions. Fertility declines and is marked in women by the menopause.

Gamete formation and fertilisation

Sexual reproduction involves production of gametes, their transfer and fertilisation. A gamete is a reproductive cell, such as a sperm or an ovum. **Gametogenesis** (gamete production) occurs in the testes and ovaries. **Spermatogenesis** is production of sperm and **oogenesis** is production of ova (egg cells).

> The testes are held outside the body cavity, so that their temperature is 2–3°C below the normal body temperature - the optimum for development of sperm.

Testes and spermatogenesis

The two testes (singular testis) are held outside the body cavity in the scrotum.

- Each testis contains about 1000 coiled **seminiferous tubules** where sperm production occurs.
- Seminiferous tubules contain sperm cells at different stages of development.
- Seminiferous tubules join to form **vasa efferentia**, tubes that transfer sperm from the testis to the **epididymis**, where they are stored.
- **Interstitial cells** between seminiferous tubules produce **testosterone**, a sex hormone.

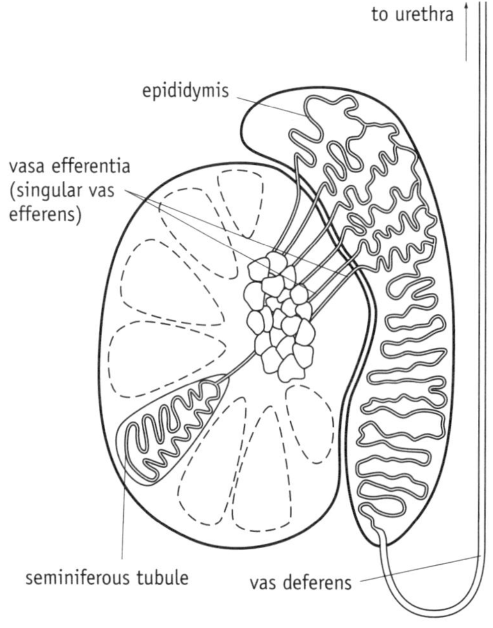

Structure of the testis

✓ Quick check 1

Spermatogenesis

Spermatogenesis starts at puberty and involves:

- **mitosis** – division of diploid ($2n$) **spermatogonia** cells of the germinal epithelium lining seminiferous tubules,
- **growth** – some spermatogonia grow into **primary spermatocytes** ($2n$),
- **meiosis** – each primary spermatocyte produces two haploid (n) **secondary spermatocytes** after the first meiotic division and **four haploid** (n) **spermatids** after the second meiotic division,
- **maturation** – while attached to Sertoli cells, spermatids lose most of their cytoplasm, form the acrosome and flagellum (tail) – becoming spermatozoa.
- Further maturation takes place in the epididymis, where sperm become motile.
- Capacitation – sperm have to spend several hours in the female reproductive system before they finally mature to be able to fertilise an egg.

spermatid
Sertoli cell
tubule lumen
secondary spermatocyte
spermatozoon
germinal epithelium
spermatogonium
primary spermatocyte

Section through a seminiferous tubule

HB

Human spermatozoa

The structure of a sperm cell consists of:

- **head** – with a haploid nucleus and **acrosome**; a vesicle containing **digestive enzymes**.
- **middle piece** – containing the axial filament and mitochondria which provide ATP for movement,
- **tail/flagellum** – to move the sperm by whip-like movements.

Ⓢ Sperm contain one copy of each gene and different combinations of alleles – because of meiosis, crossing over and independent assortment. The genotype is different from other body cells – so they would be 'seen' as foreign by the immune system.

During development sperm obtain their nutrients from **Sertoli** cells in the seminiferous tubules.

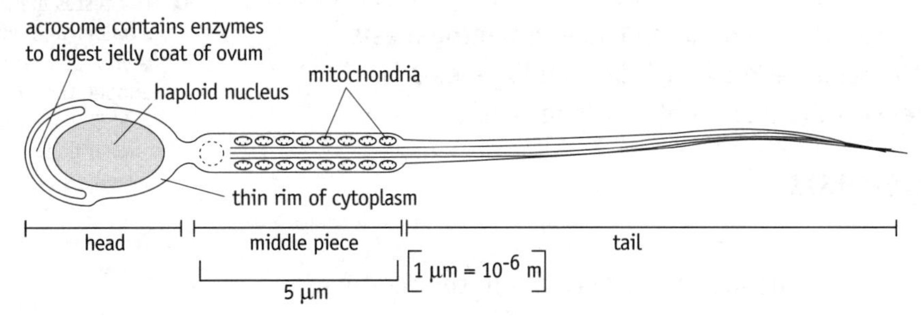

A human sperm

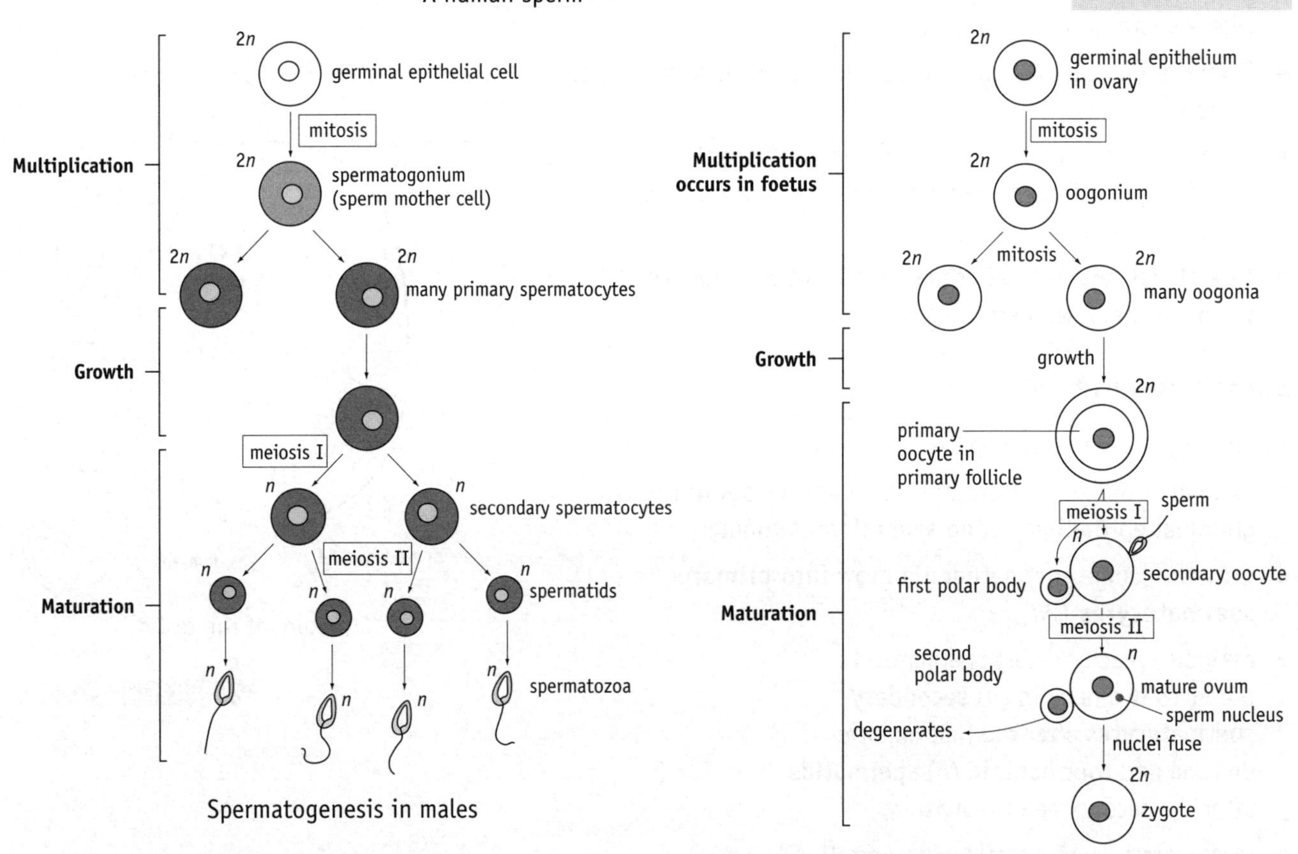

Oogenesis in females

Differences between gamete formation in males and females		
	Egg production	**Sperm production**
When mitosis of gamete forming cells takes place	before birth	from puberty to old age
When meiosis occurs	first division before birth – second at fertilisation of egg	continually
Products of meiosis	one haploid egg and two haploid polar bodies	four haploid spermatids
When gametes are produced	each month from puberty to menopause	continually from puberty to old age
Number of gametes produced	one per month	millions per day

HB

Ovaries and oogenesis

There are two ovaries. At birth each ovary consists of:

- an outer **germinal epithelium**,
- primary follicles – of which about 450 may develop into ova,
- the **stroma** or matrix, consisting of connective tissue and blood vessels.

Oogenesis

Production of ova (**oogenesis**) involves mitosis, meiosis, growth and maturation.

During embryonic development of the female;

- the **germinal epithelium** divides by **mitosis – multiplication**, producing diploid (2*n*) **oogonia**.
- oogonia **grow** and start the **first division** of **meiosis** which stops at prophase I, forming **primary oocytes** (2*n*) inside **primary follicles**.

After puberty;

- one primary follicle develops each month, **growing** into a **secondary follicle**,
- where the primary oocyte develops, completing the **first meiotic division** to form one haploid (*n*) **secondary oocyte** and one **polar body**,
- if fertilisation occurs, the **second meiotic division** is completed, producing one haploid (*n*) **ovum** and a **second polar body**.

The single ovum contains all of the cytoplasm of the original oogonium, with food reserves needed for early development of the embryo formed if the egg is fertilised.

> For both sperm and egg production, get clear in your memory when and where mitosis, growth, meiosis and maturation take place.

Fertilisation

This is fusion of a sperm and an egg to produce a diploid zygote.

- Sperm cells are transferred into the female and swim through the reproductive tract – fertilisation **takes place in the oviduct**/fallopian tube.
- Contact between a sperm and oocyte causes the **acrosome reaction**
- The acrosome releases enzymes – digest protective layers around the oocyte.
- When the cell surface membranes of the sperm and egg trouch, they fuse and the sperm nucleus enters the egg.
- This stimulates the second meiotic division of the oocyte.
- A **fertilisation membrane** develops preventing entry of other sperm.
- The sperm nucleus fuses with the ovum nucleus to produce a diploid zygote.

✓ *Quick check 2,3*

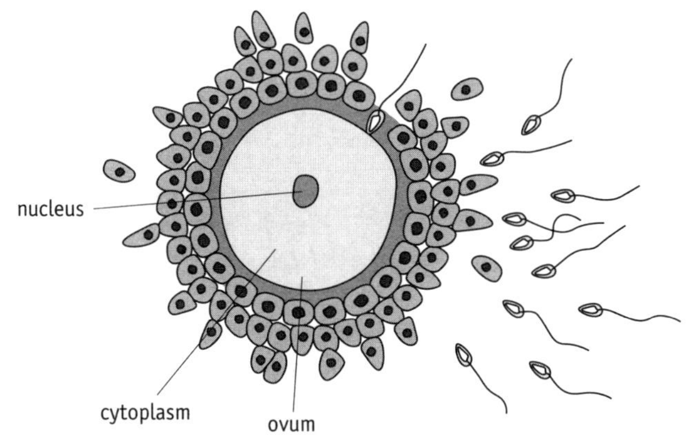

Fertilisation

> Ⓢ The acrosome contains a protease, which digests (hydrolyses) large biological molecules forming layers that prevent the sperm reaching the egg. The small, soluble products of digestion are no longer a barrier.

? *Quick check questions*

1. Describe the main stages involved in the production of sperm.
2. Give one difference between gametogenesis in males and females
3. What is the acrosome reaction?

HB

Implantation and the fetus

Implantation and early development

- The zygote produced at fertilisation travels down the oviduct to the uterus.
- It divides rapidly by **mitosis** to produce a ball of cells which becomes hollow, forming a **blastocyst**.
- The inner cell mass of the blastocyst becomes the **fetus**, the surrounding **chorion** becomes part of the placenta.
- **Implantation** – the blastocyst attaches to the endometrium/uterus wall.
- Blastocyst cells secrete enzymes that digest a way into the endometrium.
- Outer blastocyst cells secrete a hormone, **human chorionic gonadotrophin (HCG)**.
- The blastocyst develops into an embryo and then a fetus.

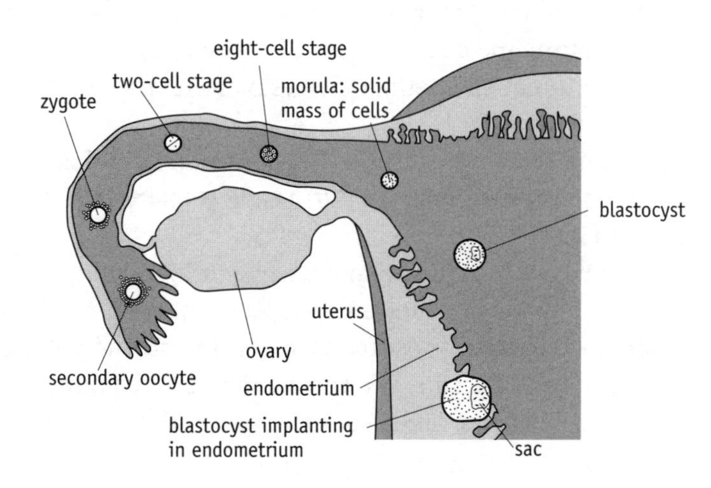

Formation of the blastocyst and implantation

✓ Quick check 1

The structure and functions of the placenta

The placenta develops from the chorion of the blastocyst and the lining of the uterus. It exchanges substances between maternal and fetal blood supplies **which do not mix.** The placenta is adapted for exchange by having;

- **a large surface area** – provided by **chorionic villi** and **microvilli** for diffusion and active transport,
- **a short diffusion pathway** – few cell layers between maternal blood and fetal capillaries,
- **many mitochondria** in epithelial cells – providing ATP for active transport,
- **a counter-current system** (maternal and fetal blood flow in opposite directions) – maintaining a **large concentration gradient** for diffusion.

There is a good blood supply to the uterus – bringing lots of oxygen and glucose for fetal respiration. Fetal haemoglobin has a higher affinity for oxygen than adult haemoglobin.

The placenta has endocrine functions, secreting **progesterone** and **oestrogen**.

The microvilli are folds of the cell membrane of epithelial cells lining the chorionic villi. The surface area of the placenta is large relative to the volume of the fetus.

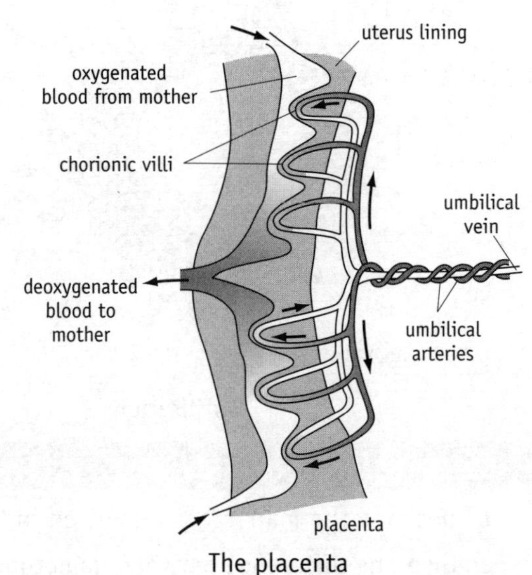

The placenta

✓ Quick check 2

HB

Exchange of materials across the placenta

Many substances diffuse across the placenta but some require active transport.

Transport of substances across the placenta	
From mother to fetus	**From fetus to mother**
oxygen glucose, lipids, fatty acids and glycerol amino acids by active transport vitamins and ions such as sodium potassium, iron and calcium by active transport antibodies viruses, alcohol, nicotine and many drugs	carbon dioxide urea other metabolic waste products

Ⓢ The cell surface membranes of cells with microvilli have many specific carrier/active transport proteins. These are specific, because of their unique tertiary structures.

Ⓢ The mother and fetus have different genotypes and their cells show different antigens. If maternal blood was able to get through the placenta, there would be a maternal immune response to fetal antigens – probably resulting in the death of the fetus.

The fetal circulatory system

The circulatory system of the fetus is different, because the placenta carries out functions usually carried out by exchange surfaces of the lungs, kidneys and gut.

Umbilical arteries and veins

- Are contained in the umbilical cord connecting the fetus to the placenta.
- The **umbilical artery** transports **deoxygenated** blood, with a high concentration of metabolic waste, from the fetus to the placenta.
- The **umbilical vein** transports **oxygenated blood** with a high concentration of nutrients from the placenta to the fetus.
- This blood **flows to the right atrium of the fetal heart** via the vena cava.
- At birth the umbilical circulation is closed off as the vessels constrict.

✓ Quick check 3

Ductus arteriosus

- This connects the pulmonary artery to the aorta, so most of the oxygenated blood flows from the pulmonary artery into the aorta.
- This bypasses the non-functioning lungs and prevents damage to the developing tissue by high blood pressure.
- Sufficient blood flows to the lung tissue to allow growth.

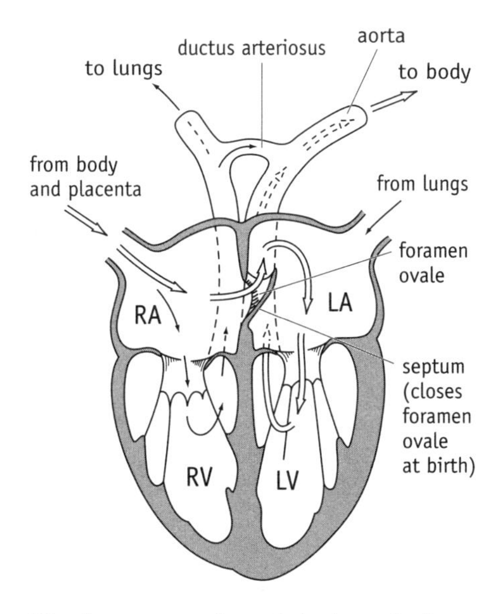

The foramen ovale and ductus arteriosus

Foramen ovale

- This is an opening between the right and left atria.
- It allows oxygenated blood to pass directly into the left atrium and then the left ventricle (and not the right ventricle).
- This reduces blood flow to the non-functioning lungs.

Ⓢ You would normally think of arteries carrying oxygenated blood and veins carrying deoxygenated blood. The situation in the umbilical cord is similar to the pulmonary arteries and veins – connecting the heart and lungs.

Following birth closure of the foramen ovale, and ductus arteriosus occurs.

Ⓢ The right ventricle pumps blood to the lungs. The left ventricle pumps blood to the rest of the body.

Quick check questions

1. What is implantation?
2. Explain two ways in which the placenta is adapted to its function.
3. Explain the function of the ductus arteriosus in the fetal circulation.

HB

Changes during pregnancy

Physiological changes in the mother

These meet the demands of the fetus and increased growth of maternal tissues.

Changes in blood volume and cardiac output

- **Cardiac output increase** by 30–40%.
- **Blood plasma volume rises** (up to 50%).

These changes increase:

- delivery of oxygen and nutrients to the placenta and removal of waste products,
- blood flow through organs of the mother; e.g. the kidneys to remove metabolic waste.
- urine production – leading to thirst and more water consumption.
- numbers of red blood cells rise with increase in blood volume, increasing the efficiency of uptake and transport of oxygen.
- More calcium ions are released into maternal blood – for fetal bone development.
- More glucose and fatty acids are present in maternal blood – for fetal respiration.

Ⓢ Cardiac output is the product of heart rate and stroke volume.

✓ Quick check 1

Thermal balance

Thermal balance is achieved when rate of heat loss equals rate of heat gain.
Thermoregulation involves controlling heat losses and gains, to maintain a constant (optimum) body temperature. A high body temperature can harm fetal development.

- The growing fetus has a high rate of respiration, producing a lot of heat.
- Excess heat is transferred to the maternal blood in the placenta.
- This heat is lost from the mother's skin by increasing blood flow to the skin (vasodilation) and increased sweating.
- Plenty of fluids should be consumed to avoid dehydration and assist cooling.

Hormonal changes during pregnancy

Changes in secretion of hormones during pregnancy begin after implantation.

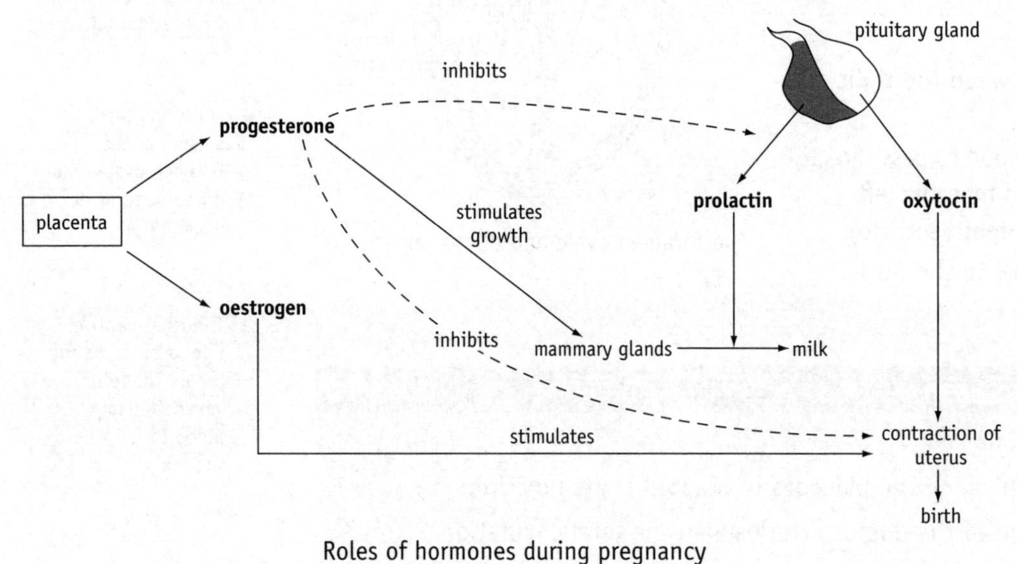

Roles of hormones during pregnancy

Human chorionic gonadotrophin (HCG)

- This is secreted by outer cells of the blastocyst which will form the placenta.
- HCG maintains the corpus luteum in the ovary so that progesterone and oestrogen secretion continues and menstruation does not occur.

Pregnancy tests use monoclonal antibody that binds to HCG in the urine of a pregnant woman,

Progesterone

For the first few weeks, progesterone is secreted by the corpus luteum. The placenta then takes over. Progesterone levels fall just before birth; one factor triggering labour. Progesterone:

- maintains thickening of the uterine lining (endometrium) – preventing menstruation that would destroy the fetus,
- inhibits release of follicle stimulating hormone (FSH) – preventing follicle and oocyte development during pregnancy,
- stimulates development of mammary glands (but not milk production),
- inhibits contraction of the smooth muscle of the uterus,
- inhibits secretion of hormones prolactin and oxytocin.

Inhibition of FSH (and luteinising hormone, LH) prevents ovulation and fertilisation during pregnancy – so no more embryos are formed or implant. More generally, the menstrual cycle stops during pregnancy.

✓ Quick check 2

Oestrogen

Oestrogen is secreted from the corpus luteum and placenta, initially in small amounts, then increasing towards birth. Oestrogen during pregnancy:

- stimulates growth of the mammary ducts,
- inhibits secretion of prolactin,
- promotes uterine muscle contraction by sensitising the uterus to oxytocin.

Prolactin

Prolactin is secreted from the anterior pituitary gland.

- During pregnancy its secretion is inhibited by progesterone and oestrogen.
- After birth, it stimulates production and release of milk by the mammary glands.
- **Lactation** involves stimulation of milk secretion by prolactin and milk ejection by oxytocin.

The loss of the placenta at birth removes the source of progesterone and oestrogen. Milk flow occurs as prolactin secretion is no longer inhibited.

✓ Quick check 3

Oxytocin

Oxytocin is secreted from the posterior pituitary gland towards the end of pregnancy.

- During early pregnancy its secretion is inhibited by progesterone.
- During birth its release stimulates contraction of the uterus muscles.
- **Positive feedback** – contractions stimulate **increased** secretion of oxytocin, leading to more frequent and stronger contractions – labour.
- Oestrogen sensitises the uterus to oxytocin.
- Following birth oxytocin stimulates release of milk from the nipple – the milk ejection reflex

Quick check questions

1. Explain the significance during pregnancy of the increase in maternal cardiac output.
2. Explain how progesterone prevents **a**) the development of eggs during pregnancy **b**) menstruation.
3. Explain the roles of prolactin and oxytocin in lactation.

The fall in progesterone and rise in oestrogen level towards the end of pregnancy enables the release of oxytocin - stimulating uterine muscle contractions resulting in birth.

HB

Growth and development

Growth is a permanent and irreversible increase in size. It can be measured by an increase in a parameter (variable) such as height or mass. **Development** is a change in structure or function.

Growth and its measurement

Possible measurements include; supine length, standing height and body mass.

- **Supine length** is measured lying down – to get the true length of the body, without the shortening effect of the body's own weight due to gravity.

Standing height should be measured at the same time of the day. People get shorter during the day, because of pressure on the discs between the vertebrae in the back!

Absolute growth

- This is the increase in a parameter over a period of time.
- It can be plotted as size, e.g. height or mass, against time.
- The graph shows the growth pattern and extent of growth of an organism.

Growth rate

- This is the change in height, length or mass in a given time period.
- It can be plotted as growth rate against time.
- The growth rate is the gradient of the absolute growth curve at a particular time.
- So, the fastest growth rate corresponds to the steepest part(s) of the absolute growth curve.

A sigmoid absolute growth curve will produce a bell-shaped absolute growth rate curve.

✓ Quick check 1

Ⓢ Phenotypic variation in height of a population of children of the same age shows continuous variation. This is the result of the influence of many genes (polygenic) and the environment

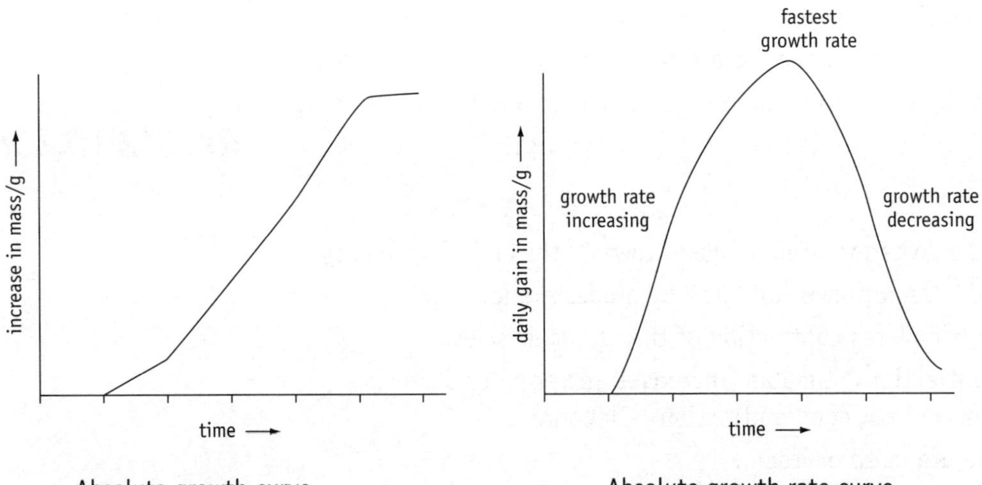

Absolute growth curve

Absolute growth rate curve

Longitudinal and cross-sectional studies

If you want to measure the growth of a population of children, as measured by increase in standing height, there are two approaches.

- **Longitudinal study** – select a group of babies at birth and measure the height of each child at regular intervals (perhaps yearly) as they grow.
- This takes a long time but produces accurate data.
- **Cross-sectional study** – select a large population of children with a range of ages from birth onwards.

- Measure heights of all children of a particular age (in years) and then repeat this for each age in the population.
- This produces results quickly but assumes that each age group's phenotype (height) is influenced by the same genotypic and environmental factors – which may not be true!

Growth in humans

Following birth, growth in humans can be divided into four phases: infancy, childhood, adolescence and adulthood.

- The growth rate is greatest during infancy and adolescence.
- A distinct growth spurt occurs in adolescence, associated with development of reproductive organs and secondary sexual characteristics.
- In adulthood growth declines.
- Males and females have similar growth patterns but the rate and timing of growth differs.
- The growth spurt usually occurs earlier in females.
- Males usually reach a greater maximum size during adulthood.

Relative rates of growth

Allometric growth – many tissues and organs grow at different rates.

- Differences in growth rate of the various organs change the proportions of the body.
- During infancy the head and brain show rapid growth.
- As humans grow, the head becomes smaller in proportion as other body parts grow faster.
- Lymphoid tissue, important in combating infection grows rapidly during childhood.
- The reproductive organs are the last organs to fully develop.

Ⓢ Young childrens' immune systems are meeting pathogenic organisms for the first time – and the antigens they carry. They are also given vaccines with antigens. Their immune systems have to react from an early age and develop memory cells.

✓ Quick check 2, 3

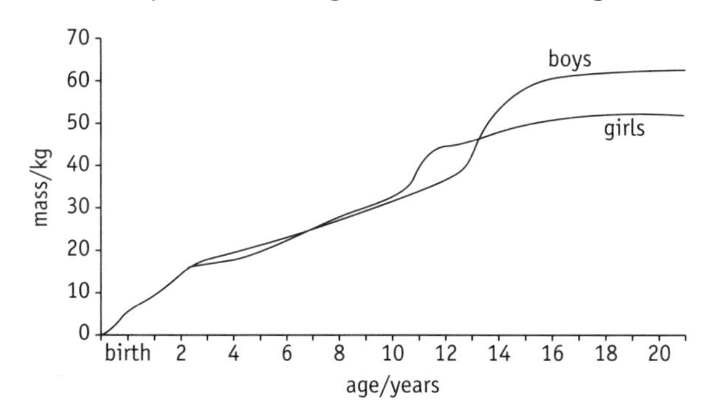

Growth curve for boys and girls

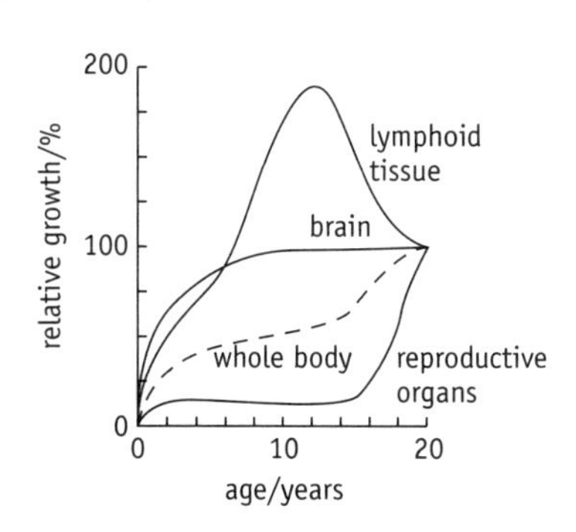

Relative growth rate curves

Quick check questions

1. Describe how the absolute growth rate could be measured.
2. Explain the advantages and disadvantages of a cross-sectional study of growth in mass of humans.
3. What causes the body proportions of different organs to vary during growth?

HB

Growth and puberty

From birth, growth is controlled by hormones. At puberty, additional hormones are produced which cause the reproductive organs to develop. These organs produce hormones which cause development of adult (secondary sexual) characteristics and gamete formation.

The **pituitary gland** is a major source of hormones affecting growth and development. It secretes:

- pituitary growth hormone (PGH),
- thyroid stimulating hormone (TSH).

Too much growth hormone during childhood produces gigantism. Too little is one cause of dwarfism.

Growth hormone

PGH is produced in large amounts during childhood and adolescence, when it stimulates:

- **protein synthesis**,
- (leading to) growth of soft tissues in the body,
- cell division by **mitosis**,
- (leading to) elongation of long **bones** and deposition of calcium in bones.

Some growth hormone is secreted in adults and controls uptake of amino acids by cells.

Ⓢ Cell growth and division needs many different types of proteins. Enzymes, are needed for DNA replication. Proteins form part of the fluid mosaic structure of membranes. There are many other examples

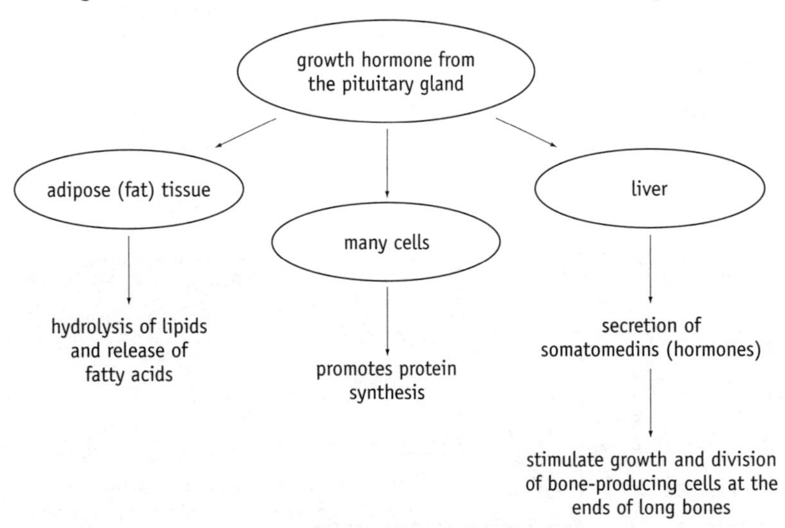

Effects of pituitary growth hormone

Thyroxine

Thyroid stimulating hormone (TSH) stimulates the **thyroid gland** to release the hormone **thyroxine.** Thyroxine **increases the rate** of **respiration** in most cells and this:

- increases the **basal metabolic rate** (BMR),
- increases heat production by cells,
- **stimulates protein synthesis.**

Because of stimulation of protein synthesis thyroxine:

- **increases growth rate,**
- is vital to normal **development** of the **skeleton** and **central nervous system** (brain and spinal cord).

Thyroxine is most important in brain development from sixth months before birth until six months after birth. Too little can produce cretinism – severe mental retardation

Secretion of thyroxine is regulated by **negative feedback**; a rise in thyroxine in the blood inhibits secretion of TSH from the pituitary gland.

✓ Quick check 1, 2

Puberty

Puberty occurs during adolescence and is associated with hormonal and physical changes; it starts earlier in girls. The **pituitary gland** starts secreting **gonadotrophins, follicle stimulating hormone (FSH)** and (later) **luteinising hormone (LH)**. These hormones act on testes of **boys** and ovaries of **girls** to:

- maintain their structure,
- stimulate gametogenesis,
- stimulate sex hormone secretion.

Gonadotrophin literally means 'feed the reproductive organs'.

In females FSH stimulates;

- development of a primary follicle each month, leading to production of a secondary follicle and secondary oocyte,
- **oestrogen** secretion from the ovaries.

A very common mistake by students is to confuse the roles of growth hormone, thyroxine and gonadotrophins. If a question asks about Growth Hormone, they write about changes in puberty and get no marks!

Oestrogen stimulates development of **secondary sexual characteristics:**

- breast development,
- growth of pubic hair,
- broadening of the pelvis.

In females LH stimulates:

- ovulation,
- development of the **corpus luteum**, which then produces the hormone **progesterone** – vital in pregnancy.

In males LH:

- stimulates **testosterone** secretion by **interstitial cells** in the testes.

Testosterone stimulates the development of secondary sexual characteristics:

- enlargement of the penis and testes,
- facial and body hair,
- growth of the larynx and deepening of voice,
- muscle and bone growth.

In males FSH:

- with testosterone stimulates spermatogenesis in the testes;

Ⓢ FSH and LH have different effects in boys and girls, because different cells have receptors for these hormones. Their cells also have different receptors. This leads to different genes being 'switched on' in the cells and different proteins produced.

✓ Quick check 3

Quick check questions

1. Explain why lack of growth hormone can cause someone to have short arms and legs.
2. Suggest why an adult who produced too much thyroxine is likely to lose weight.
3. Explain how the development of secondary sexual characteristics is controlled by pituitary gonadotrophins.

HB

Principal nutrients in the diet

In addition to water and fibre, the diet should contain sufficient quantities and correct proportions of the following principal nutrients: carbohydrates, lipids, proteins, vitamins and inorganic ions.

Roles of carbohydrates in the body

- mainly used in **respiration** providing about 17 kJ per gram when fully oxidised.
- Glucose is 'blood sugar' – carried to cells for respiration.
- Excess glucose is stored as **glycogen** in the **liver** and **muscles**.

Sugars are part of the structure of substances:

- nucleotides – deoxyribose in DNA and ribose in RNA – also part of the structure of ATP,
- many proteins are actually glycoproteins – proteins with sugars attached.

Ⓢ Glycogen is insoluble and does not affect the water potential of cells. Too much glucose in cells would make the water potential more negative. Water would enter by osmosis, possibly damaging the cells.

Roles of lipids in the body

- A phospholipid bilayer forms the basic structure of all cell membranes.
- In respiration, lipids give twice as much energy per gram as carbohydrates.
- Fatty acids are the main respiratory substrate of muscle at rest,
- Fat is the main energy storage substance of the body.
- Cholesterol gives stability to membranes and is used to make steroid hormones – such as oestrogen, testosterone and progesterone.
- Adipose tissue under the skin can provide heat insulation.
- The myelin sheath of neurones consists of lipid.

Ⓢ Cell membranes are described as a fluid mosaic – a liquid phospholipid bilayer with proteins embedded in it.

Role of proteins in the body

- Are essential for growth, repair and replacement of tissues.
- Proteins are present in all membranes in cells.
- **Fibrous** proteins have structural functions, **e.g. keratin** in nails, **collagen** in bone
- **Globular** proteins function as enzymes, antibodies and hormones, e.g. insulin.
- Proteins consist of monomers called **amino acids**.

Proteins are only used in respiration when carbohydrates and lipids are lacking in the diet.

✓ Quick check 1

Essential and non-essential amino acids

- There are 20 different amino acids commonly found in living organisms.
- Essential amino acids must be obtained in the diet they cannot be produced in the body
- Non-essential amino acids can be formed in the liver from essential amino acids by **transamination.**
- Transamination involves the transfer of the amino group (NH^{2-}) from an amino acid to an organic acid.
- e.g. the amino acid glutamic acid can be synthesised from alanine by transamination.

Alanine	+	ketoglutaric acid	$\xrightarrow{\text{transamination}}$	glutamic acid	+	pyruvic acid
essential amino acid		organic acid		non-essential amino acid		organic acid

Vitamins and inorganic ions

Vitamins and inorganic ions are required in small amounts for good health. Insufficient quantities in the diet can cause **deficiency diseases.**

- Vitamin D is needed for absorption of calcium and phosphate and bone and tooth formation.
- Calcium is needed for synthesis of bones and teeth, blood clotting and muscle contraction.
- Iron is part of haemoglobin and myoglobin molecules; lack of iron results in **anaemia**.

Deficiency of vitamin D can lead to rickets; deformation in bones – especially the legs. The bones are soft and become bent, due to the weight of the body.

Quick check 2

Food tests

Benedict's test for glucose (reducing sugar)

- Place sample in a test tube with 2cm^3 of Benedict's solution.
- Heat in a boiling water bath for 5 minutes.
- **Brick red/orange** precipitate is a +ve result.
- If the Benedict's solution remains **blue – no reducing sugar** is present.

Test for a non-reducing sugar, e.g. sucrose.

- Carry out Benedict's test on a sample to confirm a negative result.
- Hydrolyse a second sample by heating with dilute acid (e.g. HCl) or using **sucrase** (invertase) at its optimum temperature.
- Cool and neutralise with dilute sodium hydroxides solution (NaOH).
- Add Benedict's solution to test for reducing sugars (glucose and fructose) formed by hydrolysis.
- A +ve test shows a **non-reducing sugar** (sucrose) was originally present.

Foods obtained from plants usually test positive for starch

Iodine test for starch

- Add 2–3 drops of iodine solution (potassium iodine in solution).
- A **blue/black colour** indicates starch present.
- If no starch is present, the iodine solution remains **orange/yellow.**

Learn the emulsion test for lipids, as other tests for lipids may not gain marks in the exam.

Ⓢ Make sure that you know the basic structures of biological molecules and how they are joined in condensation reactions to make polymers.

Emulsion test for lipids

- Place a sample in a test tube with 2cm^3 of **ethanol.**
- Shake to dissolve any fat.
- Add water and shake the test tube.
- A **white, cloudy emulsion** of fat droplets indicates fat is present.

Quick check 3

Biuret test for proteins

- Place a sample in a test tube with 2cm^3 of **dilute sodium hydroxide** solution.
- **Dilute copper sulphate solution** (1–5%) is then added drop by drop.
- **Purple, lilac or mauve** colour indicates **protein is present.**
- If the solution remains blue, no protein is present

? Quick check questions

1 What are essential amino acids?

2 What are the roles of vitamin D and iron in the diet?

3 How would you test a food sample for the presence of lipid?

HB

Dietary requirements

Dietary requirements differ according to age, body size, sex and occupation.

Basal metabolic rate (BMR)

The BMR is the rate of **metabolism** (chemical reactions in the body) **at rest**. This uses the largest proportion of the energy content of the diet and largely determines rates at which oxygen and food are used.

- BMR varies with age, sex, size and health of an individual.

Differences in BMR – comparisons per unit mass	
Higher BMR	**Lower BMR**
men	women
children	adults
adults	elderly people
pregnant women – BMR of fetus	non-pregnant women
larger surface area:volume ratio	smaller surface area:volume ratio

- Per unit of body mass, BMR decreases with increase in body mass.
- Increase in body size usually decreases the surface area:volume ratio.
- This reduces heat loss per gram of tissue and energy requirements.

Ⓢ Lipid, protein and carbohydrate are respiratory substrates and have characteristic respiratory quotient (RQ) values; depending on how much oxygen is needed per gram for their oxidation.

Determining the energy content of food (using a calorimeter)

- A known mass of the food substance is fully burnt in a crucible.
- To heat a known volume of water.
- The increase in the temperature of the water is recorded.

energy released per gram of food / kJ = rise in temperature of the water × volume of water × 4.2

Possible sources of error include:

- heat lost to gases and body of calorimeter,
- food not completely burnt.

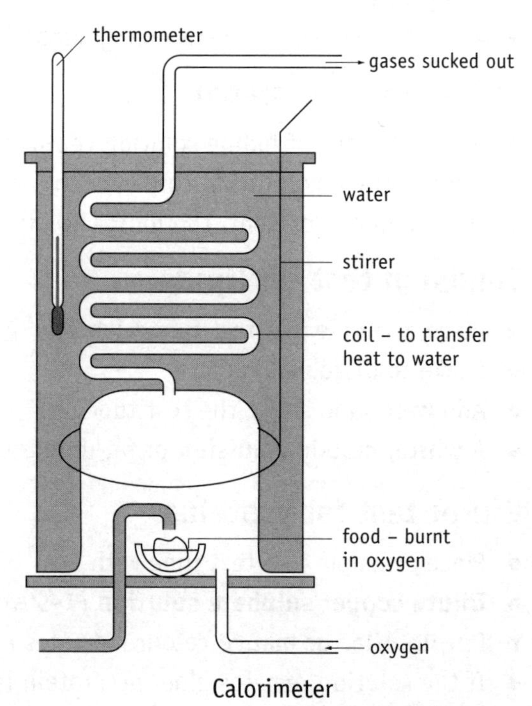

Calorimeter

✓ Quick check 1

Many books still use the old unit of energy the calorie – where the calorimeter gets its name. One kilocalorie equals 4.2 kilojoules.

Energy expenditure and protein requirements

Energy is needed for growth. Protein in the diet provides amino acids for making proteins needed for growth.

- Children have a **growth spurt** from birth to 5 or 6 years.
- A second growth spurt occurs during adolescence- associated with **puberty**.
- Girls start puberty earlier than boys.
- Each spurt increases energy expenditure and the need for protein in the diet.
- Adults are not growing, so their energy and protein needs are less.
- **Ageing** involves loss of tissue and (usually) a less active life style – so elderly peoples' energy and protein needs are less.

- At any age the more physically active a person is, the higher their energy needs.
- An increase in exercise can cause muscle growth – requiring more protein.
- Athletes eat large amounts of carbohydrate in the days before a event -maximising glycogen storage in muscle and liver cells.
- Glycogen can be rapidly converted to glucose to be used in respiration, providing a quick source of energy – enhancing athletic performance.

Energy needs for men, women and children

✓ Quick check 2

Balanced diet

A balanced diet has the correct quantities and proportions of carbohydrates, lipids, proteins, vitamins, inorganic ions, water and fibre required to maintain health.

Weight loss diets may lead to:

- use of muscle and other tissue protein for energy production – causing wasting,
- increased risk of infection, vitamin and mineral deficiency diseases,
- tooth decay, low blood pressure, constipation, menstruation ceasing.

Vegetarian diets:

- must be varied to supply all the essential amino acids;
- may yield lower total energy values;
- result in reduced intake of vitamin B_{12}, vitamin D, iron, zinc and iodine;
- can result in anaemia if insufficient iron is included.

Dietary demands of pregnancy and lactation

Requirement	Reason for increased dietary demand	
	Pregnancy	Lactation
energy content	growth of fetus, placenta and uterus	milk synthesis
protein	growth of fetus, placenta, uterus and breasts	high amino acid content of milk – for growth of baby
iron	for fetal haemoglobin and increase in mother's haemoglobin and blood volume	synthesis of baby's haemoglobin
calcium	growth of fetal teeth and bones	growth of baby's bones (and teeth)

Ⓢ Protein synthesis uses information from DNA via transcription and translation.

A deficiency of protein in the diet of children can cause Kwashiorkor – with stunted growth.

Recommended daily amounts (RDAs) of food energy and nutrients are available for different groups of the population depending on age, sex, activity etc.

Due to faster growth rates, the protein requirement per gram of body mass is higher in children than in adults.

Requirements for iron in females

Females may require more iron in their diet than males.

- They lose iron with menstrual blood loss.
- Intrauterine devices (IUDs) may increase menstrual loss.
- Oral contraceptive pill often produces shorter periods with less blood loss.

✓ Quick check 3

? Quick check questions

1. What is meant by the basal metabolic rate?
2. What problems can arise from weight reducing diets?
3. Suggest how glycogen loading can enhance athletic performance.

HB

Muscle structure and the sliding-filament theory of muscle contraction

Muscle contracts and generates force which can produce movement. Muscle **can only contract**, an outside force has to act to lengthen it after contraction.

Structure of skeletal muscle

Skeletal muscle is also called **striped** or **striated muscle.** Under a light microscope muscle consists of many long **muscle fibres** or cells.

Muscle fibres:

- are cylindrical in shape and enclosed by a cell surface membrane or **sarcolemma**;
- have many nuclei (**multinucleate**);
- contain numerous protein strands or **myofibrils** with characteristic cross-striations;
- are arranged in parallel, giving a striped appearance as cross-striations line up;
- are surrounded by collagen and connective tissue which extends to form the **tendon** connecting muscle to bone.

Ultrastructure of skeletal muscle

The ultrastructure of muscle can be seen using electron microscopy. The banding pattern of skeletal muscle is due to the arrangement of filaments made from **myosin** or **actin** proteins. Each repeating pattern of banding is called a **sarcomere.**

- Thin filaments are made of actin, thick filaments are made of myosin.
- Thick filaments are present in the **A-band** or **dark band.**
- The outer regions of the A-band are darker, because they contain overlapping myosin and actin filaments.
- The **H zone** at the centre of the A-band is not as dark, because it contains only myosin filaments.
- The M line connects the myosin filaments in the A-band.
- **I-Band** or **light band contains** only thin actin filaments.
- The **Z line** connects these actin filaments.

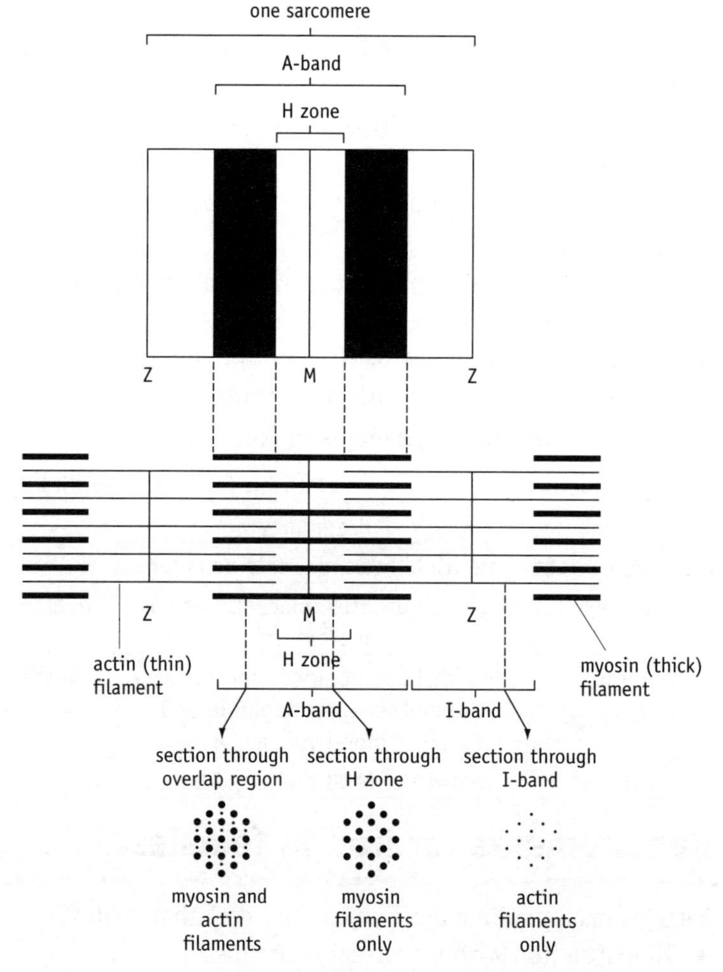

Structure of skeletal muscle

Ⓢ There are many mitochondria next to the sarcomeres. These provide the energy for contraction - ATP from respiration.

✓ *Quick check 1*

HB

Sliding-filament hypothesis

This describes the mechanism of muscle contraction. During contraction thin actin filaments are pulled past and between the thick myosin filaments. This causes shortening of the muscle fibre and changes in the banding pattern – the filaments themselves do not contract or shorten.

- The **H zone** inside the A-band **narrows** – contains only myosin filaments.
- The **outer darker regions** of the A-band **widen** – contain overlapping actin and myosin filaments.
- The **I-band** (light band) **shortens** - the non-overlapping portion of the actin filaments.
- **The** A-band **stays the** same – **the length of myosin filaments does not alter**.
- **The** Z lines **are pulled towards each other**.

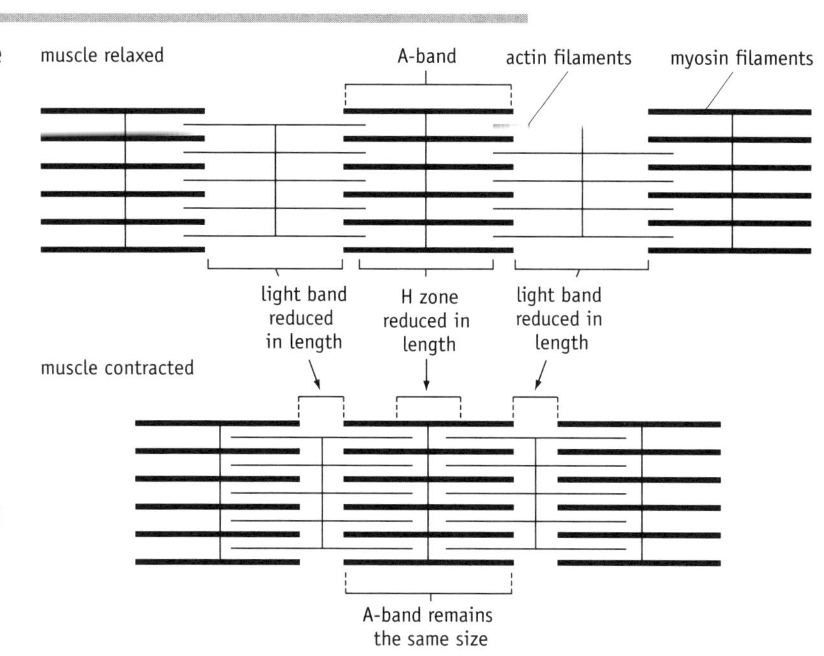

Sliding-filament hypothesis

Actin and myosin filaments do not contract, they slide past each other – pulling Z lines closer together.

Role of ATP - and the ratchet mechanism

This explains muscle contraction and involves the formation of cross bridges between actin and myosin filaments.

- Each myosin filament has a long rod-like region and a myosin 'head'.
- Actin filaments have attachment sites for the myosin heads.
- During contraction actomyosin bridges form as myosin heads attach to actin filaments.
- Bridges rapidly break and reform along the actin filaments pulling them past the myosin filaments.
- ATP provides energy for the release of myosin heads from actin.
- ATP is hydrolysed by an ATPase enzyme on the myosin heads.
- Numerous mitochondria supply ATP via aerobic respiration.

Ratchet mechanism

Ⓢ Actin and myosin are proteins – with specific tertiary structures. The myosin head fits into the site on actin, because of a 3D shape fit. When myosin binds to actin, that changes its environment – causing a shape change that produces force – producing movement.

Quick check questions?

1 During muscle contraction what happens to the length of the (**a**) A-band (**b**) H zone?

HB

Control of muscle contraction and muscles as effectors

Role of calcium ions

Muscle contraction is activated by the calcium ions released from the sarcoplasmic reticulum in the muscle fibre when the fibre is stimulated.

- The binding site on actin filaments in relaxed muscle is covered by the protein, tropomyosin.
- Tropomyosin switches on or off the contraction mechanism.
- Tropomyosin is attached to another protein, troponin.
- When released, calcium ions bind to troponin causing it and the attached tropomyosin to move from the binding site.
- The actin filament is now switched 'on' and myosin binds to form cross bridges.
- Bridges rapidly break and reform causing shortening of each sacromere.
- When the muscle is no longer stimulated the calcium ions are actively moved back into the sarcoplasmic reticulum.
- They stimulate the action of ATPase which hydrolyses ATP providing the energy for the formation and breakdown of the bridges.

Ⓢ Calcium ions are released when electrical impulses travel from a neuromuscular junction. Nerve impulses travel down the axon of a motor neurone to get to the neuromuscular junction.

Ⓢ The binding of calcium causes a change in the shape of the troponin protein – which changes its function.

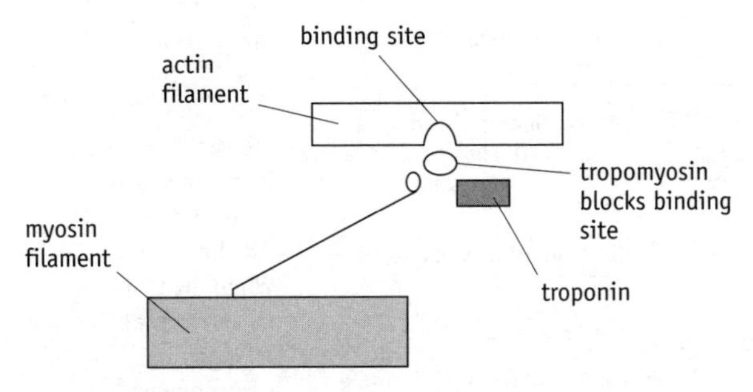

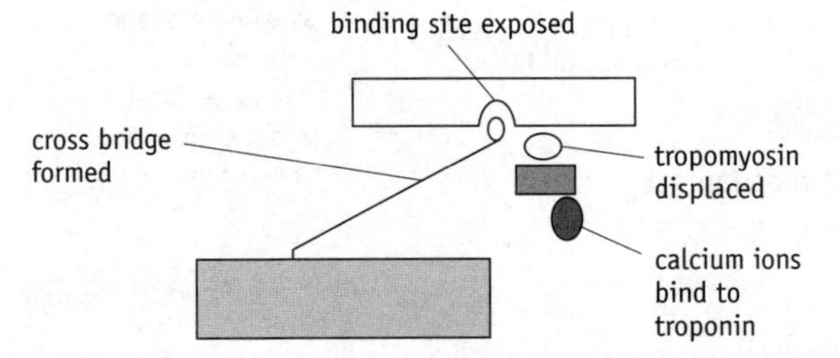

Role of calcium, troponin and tropomyosin in muscle contraction

HB

Muscles as effectors

Phosphocreatine and muscle contraction

If muscles need to contract for long periods of time, ATP is used faster than it can be supplied by respiration. Muscles need to replace ATP rapidly.

- Phosphocreatine acts as an energy store – used to regenerate ATP rapidly.
- Skeletal muscles contain about three times more creatine than ATP.
- When the muscle is resting, creatine accepts a phosphate group from ATP to form phosphocreatine.
- During periods of prolonged contraction, phosphocreatine transfers a phosphate group to ADP to form ATP.

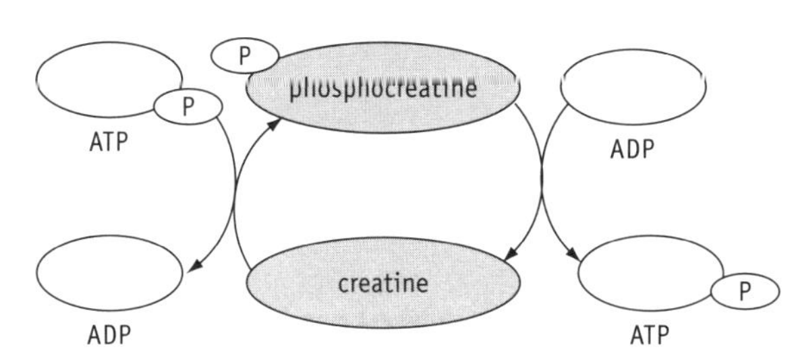

Creatine/phosphocreatine cycle

Quick check 1

Quick check 2

Slow and fast skeletal muscle fibres

Skeletal muscles contain slow muscle fibres, fast muscle fibres, or a mixture of both.

- 'Slow' and 'fast' refer to **speed of contraction** – the time to reach maximum tension.
- Slow and fast fibres have slightly different myosin molecules – with **different ATPase enzyme** on the myosin heads.

Slow fibres	Fast fibres
mainly aerobic respiration	mainly anaerobic respiration
little glycogen	large amount of glycogen
lots of capillaries	few capillaries
high myoglobin content	little myoglobin
many mitochondria	fewer mitochondria

If someone train for an endurance event, such as a long distance run, both types of fibre develop more mitochondria.

Slow fibres:

- are found in muscles involved in **posture** – many in the arms, legs and back,
- **tire (fatigue) slowly**, because they can respire aerobically – so can work for long periods of time,
- have **myoglobin** – which binds oxygen and acts as an oxygen reserve,
- are red, because of their myoglobin pigment.

Fast fibres:

- are found in muscles responsible for **fast movements**, e.g. muscles that move the eye in its socket,
- tire (fatigue) relatively quickly, because they rely on anaerobic respiration.,
- are white, because they lack myoglobin.

Some muscles have slow and fast fibres, e.g. the gastrocnemius in the leg. The slow fibres are used for normal activities, such as walking, and the fast for short bursts of fast movement, such as sprinting.

Quick check 3

Quick check questions?

1. Explain how calcium ions activate muscle contraction.
2. Explain how energy is provided for the ratchet mechanism during muscle contraction.
3. Explain why slow muscle fibres do not tire rapidly.

HB

Senescence

Body changes and deterioration of physiological functions associated with **ageing** are known as **senescence**. This phase of decline is a characteristic feature of the human life-span.

Changes in physiological effectiveness

Ageing produces declines in the effectiveness of many physiological systems.

- Cardiac muscle fibres weaken and cardiac output decreases.
- There is a decline in the speed of conduction of nerve impulses (10–15%).
- Speed of response (reflexes) slows down.
- **Basal metabolic rate decreases**, because ageing involves loss of tissue and (usually) a less active life style – so elderly peoples' energy needs are less.
- The more physically active a person is, the higher their energy needs.

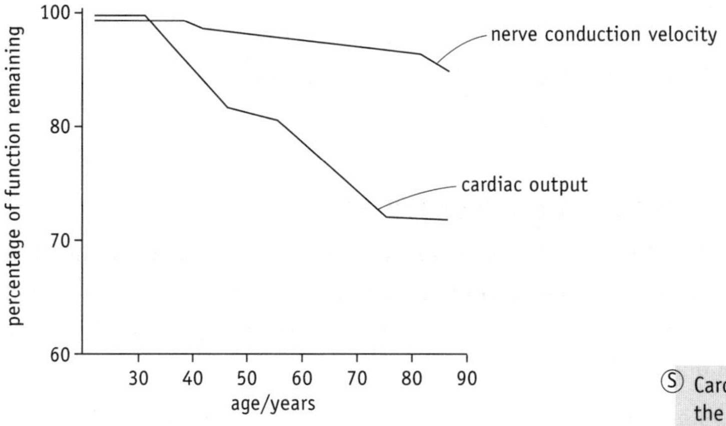

Effect of ageing

Ⓢ Cardiac output is the product of heart rate and stroke volume. It increases during exercise.

✓ Quick check 1

Changes in reproductive function in females

Sometime around the age of 50, women go through the menopause.

- Fertility reduces as the number of viable follicles in the ovaries decreases.
- Frequencies of menstruation and ovulation diminish and stop.
- Oestrogen secretion by the ovaries decreases and then stops.

Loss of oestrogen produces the symptoms of menopause:

- atrophy (loss of tissues) of the vagina, cervix, uterus, ovaries and oviducts,
- increased risk of atherosclerosis and heart disease,
- faster osteoporosis,
- hot flushes.

Pituitary gonadotrophin (FSH and LH) levels rise to a peak 2–3 years after menopause.

- Oestrogen inhibits secretion of FSH and LH.
- The fall in oestrogen at menopause removes this inhibition, so more FSH and LH are secreted.

✓ Quick check 2

? Quick check questions

1 Describe the physiological changes that take place during ageing.

2 Explain why a woman becomes infertile at menopause.

Module 7: end-of-module questions

1 The diagram shows a liver cell, its receptors for insulin and glucagon and an outline of the metabolic pathways involving glycogen.

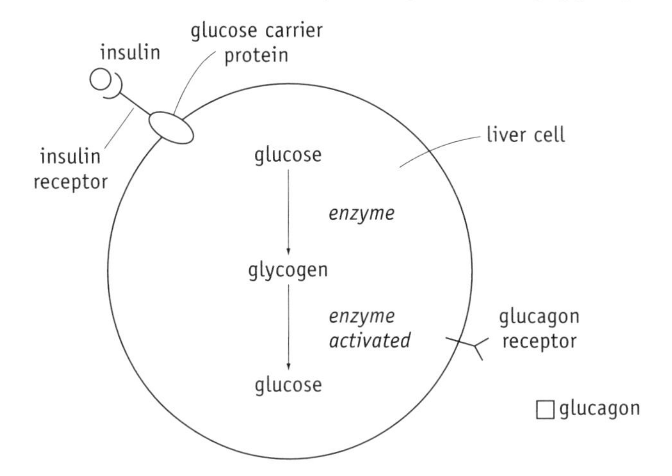

a Use information in the diagram and your own knowledge to explain:

i how the release of insulin by the pancreas leads to a reduction in glucose in the blood. [3]

ii how glucagon release leads to the release of glucose into the blood. [2]

b Some people inherit an allele which produces a faulty enzyme in the pathway between glycogen and glucose. Suggest what symptoms they would show. [2]

2 a Explain how and where proteins are digested in the human gut. [5]

b Describe how the release of pancreatic secretions is controlled. [4]

3 a Other than the control of heart rate, explain **two** differences between the sympathetic and parasympathetic nervous systems in terms of their effects on human physiology. [4]

b Explain how heart rate is increased during exercise. [4]

4 The diagram shows the events leading to the release of acetylcholine into the synaptic cleft.

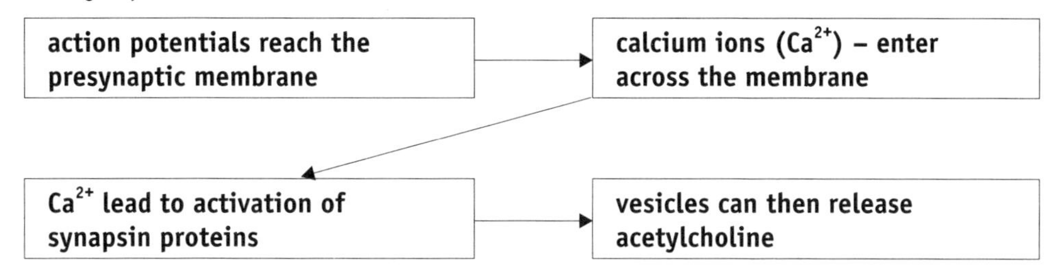

a **i** Explain how the vesicles release acetylcholine. [2]

ii Explain what happens to acetylcholine after it is released. [3]

HB

b The bacterium *Clostridium botulinum* produces a toxin which is an enzyme. This is an endopeptidase that breaks down synapsin.

i Suggest how the toxin affects humans. [3]

ii Explain what the toxin does to synapsin. [2]

c Explain what is meant by spatial summation. [2]

5 The graph shows the changes in potential across part of the membrane of the axon of a motor neurone during an action potential.

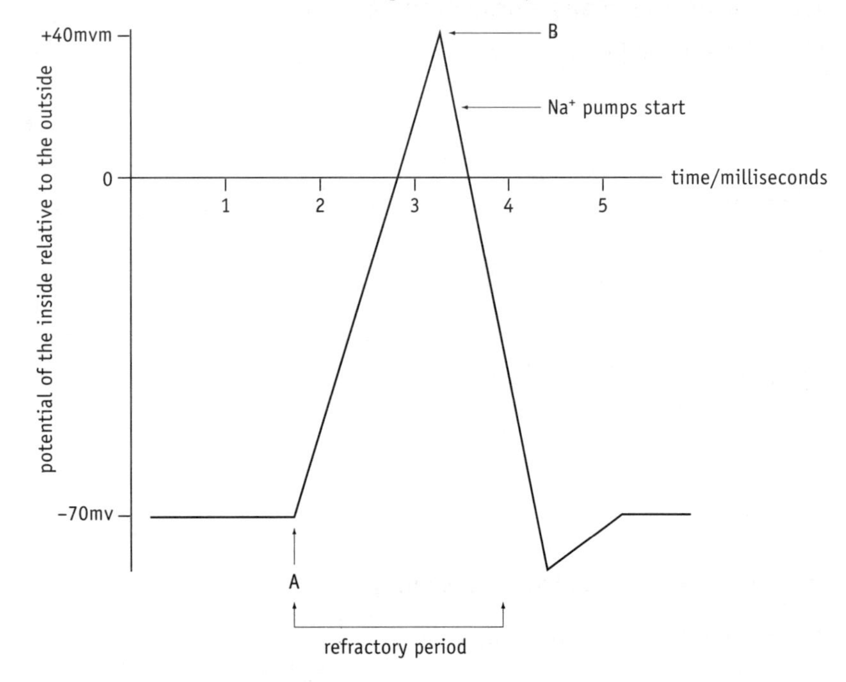

a Explain what happens at **A**. [3]

b Explain what happens at **B**. [3]

c Use the graph to calculate the maximum frequency of nerve impulses passing along this axon. [2]

6 The graph shows the average energy needs of children (both sexes) and men and women of different ages.

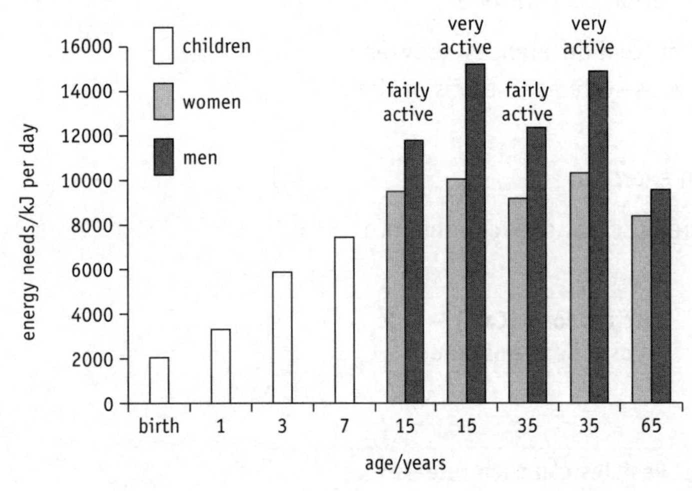

a **i** Calculate the percentage increase in energy needs between birth and three years of age. [2]

ii Suggest why children were all grouped together up to age seven but then separated into men and women from 15 onwards. [2]

b Explain the differences in energy needs between men and women aged 15 and 35 who are:

i fairly active, [1]

ii very active. [2]

c Explain **two** reasons why the energy needs of people aged 65 are lower than those aged 15 and 35. [2]

7 a Explain how a sarcomere contracts. [5]

b The graph shows the development of tension in two types of muscle fibre, **X** and **Y** and a muscle, **M**

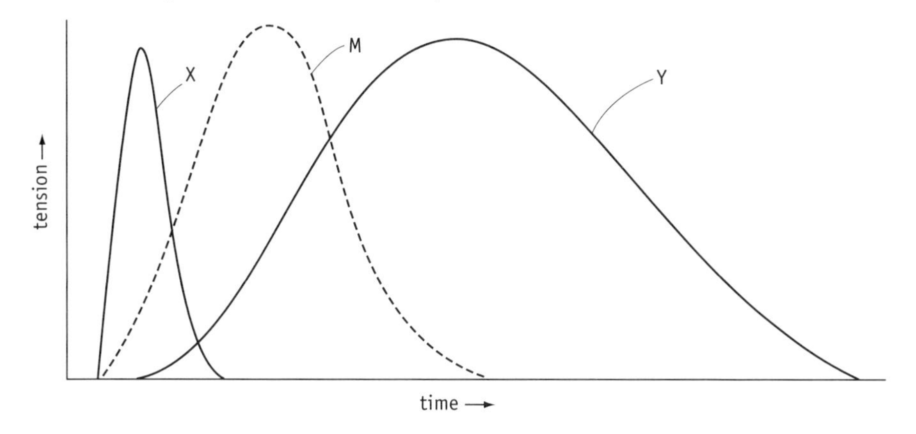

i Name the two types of muscle fibre that produce **X** and **Y**. [1]

ii Muscle **M** is a large muscle in the leg, involved in posture, walking and running. Suggest an explanation for the shape of its curve. [3]

8 a Explain **two** differences between gametogenesis in males and females. [4]

b The foramen ovale and ductus arteriosus produce a different blood flow in the fetus to that seen after birth.

i Explain the functions of the foramen ovale and ductus arteriosus. [4]

ii In some babies the foramen ovale does not close after birth. Suggest the affects this would have on the baby. [3]

Appendix: Exam Tips

Lots of marks are lost by not answering questions as they are set, or not knowing material in the syllabus. You must know the information, terms and examples that are included in the syllabus – the exam board only allows questions that can be answered using syllabus material. Other information will not harm you, but will not be necessary to get a good mark!

Describe

'Describe' means put information into words. The information is usually **given** to you in a table, graph or diagram.

Example: the graph shows the rate of reaction of an enzyme with different concentrations of substrate and how the rate is affected by the addition of a particular concentration of an inhibitor.

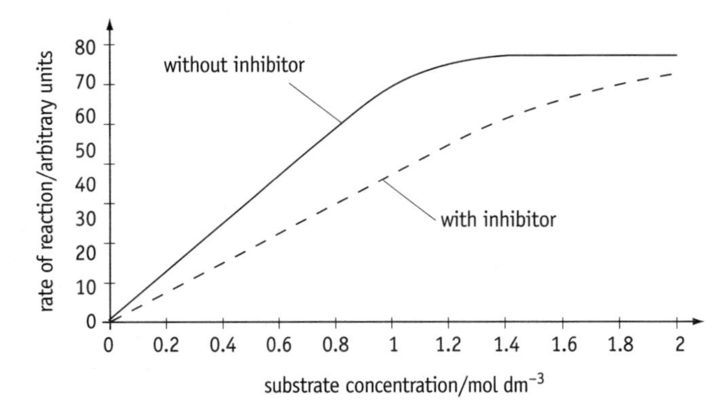

Question

Describe the effect of each of the following on the rate of reaction of the enzyme:

a the concentration of substrate;

b the inhibitor.

Answers

a Between substrate concentrations of 0 and 1 mol dm^{-3}, the rate is directly proportional to the concentration of substrate. The rate reaches a maximum at a substrate concentration of about 1.4 mol dm^{-3}.

b The inhibitor reduces the rate of action of the enzyme at lower concentrations of substrate. Above substrate concentrations of 1.2 mol dm^{-3}, the rate with the inhibitor gets closer and closer to the rate without the inhibitor.

These answers **describe** what you can see on the graph – in reasonable detail.

Explain

'Explain' means that you should 'know' the answer from syllabus material that you have been taught. The following question uses the same graph and information as for 'Describe'.

Question

Is the inhibitor competitive or non-competitive? **Explain** your answer.

Answer

The inhibitor is competitive, because its effect is overcome by increasing the concentration of substrate.

A non-competitive inhibitor would inhibit the enzyme at any concentration of substrate.

Suggest

'Suggest' means that you are unlikely to have seen the material in the question but you should have been taught things from the syllabus that will allow you to answer. The following question applies to an enzyme and its inhibitor.

Question

Two substances, **X** and **Y**, were investigated as possible rat poisons. Both inhibit the same enzyme in an important biological process. The diagrams show the structure of the enzyme, its substrate and the inhibitors. Use the information in the diagrams to suggest which inhibitor would be the best poison.

Answer

Inhibitor Y would be best, because it is non-competitive. It binds to a site other than the active site.

Its effect cannot be overcome by more substrate (unlike inhibitor Y) and so the metabolic pathway is blocked.

OR

Inhibitor X would be best, because it is competitive. It binds to the active site. This reduces/stops the substrate being turned into product (and stops the metabolic pathway).

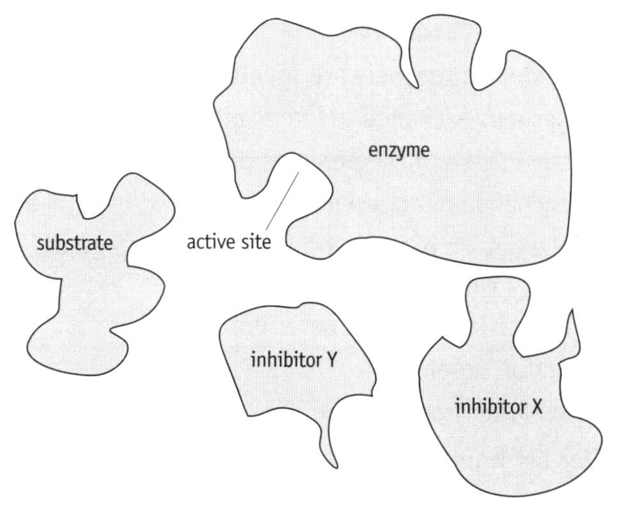

Rat poison

It is not unusual to have alternative answers to this sort of question. You are not expected to have learnt this material – it isn't specifically given in the syllabus. Either answer is a reasonable interpretation of the information.

Synoptic questions

These are part of the exams for Modules 6 or 7. There is also a complete synoptic paper. Synoptic questions require you to bring together knowledge and ideas from all the previous Modules – AS Modules 1, 2 or 3 and A2 Module 5. You are **not expected to remember all the detailed knowledge from previous modules** and you will not be asked for specific details.

Example

In a Module 5 exam you might be asked a series of questions about the roles of saprophytic bacteria and fungi in decomposition and the recycling of nutrients.

- You might then be given data showing that the rate of recycling of nutrients increases in Spring and Summer, compared to the Winter and be asked to suggest why this occurs.
- The answer involves warmer climatic conditions in Spring and Summer, leading to faster rates of action of enzymes of bacteria and fungi and their faster digestion of organic remains. This is using basic knowledge about a factor affecting enzyme activity – from Module 1.

Essays

In the **Biological Principles** exam you will have to write **one** essay. There will be a **choice from two titles**. These will ask about important ideas that apply to many areas of the specification. One example used by the exam board in training material is given below.

The process of diffusion and its importance in living organisms.

Diffusion is involved in many topics in many modules.

- The process itself is in AS Module 1 (AS1).
- Together with osmosis as a special case of diffusion of water and how substances enter and leave cells.

Diffusion or osmosis are involved in, for example:

- gaseous exchange in lungs, gills and leaves,
- the uptake of the products of digestion,
- exchange of materials between blood in capillaries and tissues,
- the uptake of water by roots and root pressure and in the mass flow hypothesis of translocation of sugars,
- regulation of blood water potential,
- action potentials,
- synaptic transmission.

This list would make a good **essay plan!**

To get a good mark you need to write about as many of these as you can. You need to include **important concepts** and the **correct terms. Do not** write at length about just one example. If you are studying Biology, **do not** write just about humans – always look for the chance to mention other organisms – especially plants!

Very few good essays will be more than **about three pages** long – there are no marks for the amount you write! You do not need an introduction and conclusion! You must write in continuous prose – no lists or bullet points.

Marks are awarded as follows:

- **16 for knowledge and understanding,**
- **3 for breadth of knowledge,**
- **3 for relevance,**
- **3 for quality of written communication.**

In our example about diffusion, if you write a paragraph about each of the things on the list, you could get all the marks for knowledge and understanding, breadth and relevance.

Answers to quick check questions

Module 5: Inheritance, Evolution and Ecosystems B HB

Meiosis (and inheritance)

1 a The genes an organism has and the alleles of those genes.

b The characteristics of an organism due to its genotype and interactions between the genotype and environment.

2 Meiosis produces haploid cells with one copy of each chromosome; when gametes fuse, the diploid number of chromosomes is restored; if diploid gametes fused, the number of chromosomes would double with each generation.

3 AbH due to: crossing over between genes A and B; independent assortment such that H rather than h went with the Ab chromosome.

Principles of Mendelian inheritance

1 Since all F_1 round, round is dominant to wrinkled; all F_1 heterozygous; expect 25% of F_2 to be homozygous and wrinkled; 3:1 ratio round to wrinkled.

2 Use *diagrams to help your explanation*. Both parents must carry O. Parental genotypes thus I^AI^O and I^BI^O . Possible genotypes of offspring I^AI^O, I^BI^O, I^OI^O, I^AI^B.

3 If neither parent affected, then mother carries haemophilia; recessive allele on X chromosome; 50% of her eggs have haemophilia allele; so 50% of sons affected (since they get a Y from father); 50% of daughters will be carriers. *You could use diagrams to help your explanation.*

Dihybrid crosses and chi-squared

1 Since F_1 all purple flowers and green pods, purple dominant to white and green dominant to yellow; F_1 heterozygous PpGg; gametes PG, Pg, pG, pg in equal numbers; Punnett square should give 9:3:3:1 ratio; 9 purple green: 3 purple yellow: 3 white green :1 white yellow.

2 a Expected results are calculated from the phenotypic ratio expected from a genetic cross and the number of offspring obtained in an experimental cross. Observed results are the numbers of each phenotype counted in the experiment.

b A significant difference is one which is too great to be due to chance alone.

Investigating variation

1 a Human masses – use a histogram, because data is continuous and quantitative.

b Human eye colour – use a bar chart, because the data is qualitative and in discrete classes.

2 Standard deviation is a measure of the spread of data about the mean. Standard error is a measure of the variation in the value of the mean.

Causes of variation

1 IQ and length of hair are continuous; there is a range betwen two extremes; it can be measured; there are no distinct groups. Haemophilia is discontinuous; it can not be measured; there are two distinct groups, those with the it and those without.

2 Blood groups are genetically determined; with little/no environmental factors; so, might easily have different blood groups for non-identical, since they have different genotypes; height and weight environmental factors important; growing in same environment these likely to be more similar.

3 Crossing over at chiasmata; produces new combinations of maternal and paternal alleles; independent assortment; produces new combinations of maternal and paternal chromosomes and their alleles.

Hardy-Weinberg principle

1 a $p^2 + 2pq + q^2 = 1$: Let p stand for frequency of the dominant tall allele and q for the short allele. Frequency of the homozygous short is 9% – a probability of 0.09 for q^2 the homozygous short plants. Square root of 0.09 = 0.3 , which is the probability for the short allele. We know that p + q = 1, so the probability of the tall allele p = 1 – 0.3 = 0.7. This means that the frequency of the tall allele is 70%.

b The probability of the heterozygous plants = 2pq = 2 x 0.3 x 0.7 = 0.42 This is a frequency of 42%. *Note, The probability of homozygous tall = 0.7^2 = 0.49 and 0.49 + 0.42 + 0.09 = 1.*

2 There is variation in genotypes and phenotypes of a population. Some phenotypes are better adapted to the environment than others and are more likely to survive; they have a higher differential survival rate. The better adapted are thus more likely to pass on their combinations of alleles to the next generation. The more poorly adapted are less likely to pass on their alleles. The alleles of the better adapted become more common – there is a change in the allele frequencies in the population – which is evolution.

3 a Attempt to breed beetles from the island with beetles from the mainland. If they fail to produce fertile offspring, then they are different species.

b The beetles on the island were geographically isolated from mainland populations. The island population adapted to its environment through natural selection. The isolation prevented immigration gene/allele flow between the mainland and island populations. This allowed the island population to evolve until its gene pool become so different that reproductive isolation developed.

Selection and change in allele frequency

1 Variation in population; some phenotypes better adapted to environment than others; compete better, have a selective advantage; produces differential survival rates; better adapted more likely to survive long enough to reproduce.

2 Answer implies that the rats decided to develop a resistance gene/allele in response to warfarin; if an allele/gene giving resistance is present, it is due to a chance mutation.

3 People homozygous for sickle-cell anaemia are not likely to survive to reproduce (in many parts of the world). This removes sickle-cell alleles from the population in each generation, leading to fewer heterozygous in the next generation. In fact, there tends to be a fairly constant frequency of heterozygotes from generation to generation.

Classification

1 *Should be written as* Homo sapiens (lower case for sapiens).

2 Humans and chimpanzees are both primates; being in the same order, they share a common ancestor; which had the genes they share.

3 Plants photosynthetic, fungi saprophytic or parasitic; plant cell walls of cellulose, fungi of chitin; plants multicellular, fungi unicellular or hyphae.

Ecosystems, ecological terms and ecological techniques

1 Set out a grid system on the lawn (using string). Use a number/letter system to identify intersections on the grid. Use a random number table to pick 20 intersects at random. Place a frame quadrat on each chosen intersection and record the number of squares in each quadrat covering dandelions. Calculate the average number of squares covering dandelions and convert to percentage cover.

2 $n_1 = 50$, the number caught and marked – $n_2 = 50$, the number caught second time – $n_m = 15$, the number marked in the second sample. Population size = (50 x 50) ÷ 15 = 167

Diversity

1 High diversity; many different species; and large numbers of individuals of species; abiotic environment which is not extreme; with respect to abiotic factors.

2 Low solar energy, few tree species able to carry out enough photosynthesis to survive; or survive cold conditions; extreme abiotic conditions; as energy input increases, more light for photosynthesis; temperatures nearer to optimum for enzymes; above certain point, temperatures become too hot for many species; abiotic conditions becoming extreme again.

3 Diversity usually measured using diversity index; this requires the number of individuals of each species, as well as the number of species.

Succession and climax communities

1 Changes in communities over time with one following another, until climax community established; often starts with colonisers on bare soil; this community makes biotic and abiotic factors less severe; especially soil factors; e.g. humus to hold water; making it possible for other plants to establish a new community; which outcompete original colonisers; repeated with later communities, until climax community – stable over long period of time.

2 Herbaceous plants change soil factors; when they die and decompose; add humus to soil which holds water; source of mineral ions such as nitrate; improve structure of soil; plants also act as wind breaks, increasing temperature locally/reducing water loss; make abiotic conditions less severe, so seedlings of woody plants can grow.

3 Interspecific competition; marram can not compete with woodland plants; seedlings can not get established.

Photosynthesis

1 Photolysis of water gives electrons to replace those lost from chlorophyll; and hydrogen ions which reduce NADP to NADPH.

2 Light energy absorbed by chlorophyll molecules; excited electrons leave chlorophyll; electrons are a source of chemical energy/reducing power.

3 Hydrogen carbonate ions; glycerate 3-phosphate; ribulose bisphosphate and glucose.

4 No light, no light-dependent stage; NADPH production stops; so soon no NADPH to reduce glycerate 3-phosphate to sugar.

Energy transfer

1 Very little energy passes from one trophic level to the next; by 5/6 trophic levels there is no enrgy to pass on/not enough to support another level.

2 93 + 6 ÷ 140 x 100 = 70.7%.

3 Pyramid of number – the number of individuals per unit area at each trophic level at a specific time. Pyramid of mass – mass (Kg) of organisms at each trophic level at a specific time and in a given area (m^2) – units of Kgm^{-2}. Pyramid of energy – energy (kJ) in each trophic level, for a particular area (m^2) and period of time (yr) – units of $kJ\ m^{-2}\ yr^{-1}$.

4 Difficult to find energy content of organisms; would destroy organisms; hard to know how much of the energy present was for that year.

Energy supply

1 All organisms respire; plants also photosynthesise.

2 ATP phosphorylates other molecules; making them more reactive; and lowering activation energy needed.

3 A cat probably eats a lot of meat – containing lipid and protein – so RQ probably between 0.7 and 0.9.

Respiration

1 (6-carbon) glucose oxidised to 2 molecules of (3-carbon) pyruvate; by removal of hydrogen; accepted by NAD to give reduced coenzyme.

2 In Krebs, 6-carbon compound oxidised to a 4-carbon compound; by removal of hydrogen; this accepted by NAD and FAD to give reduced coenzymes; supply chemical energy/reducing power for ATP production by oxidative phosphorylation.

3 Reduced coenzyme NADH/FAD reduces first protein carrier; electrons pass down carrier chain; in series of oxidation and reduction reactions; lose energy as they do; some of the energy used to make ATP.

4 Oxygen the final electron acceptor from oxidative phosphorylation; if electrons not accepted, all protein carriers soon in reduced state and no electrons flow; so no oxidative phosphorylation or ATP; glycolysis does not produce enough ATP to keep us alive.

Nutrient cycles

1 Photosynthesis is where inorganic carbon in CO_2 enters the biotic environment; respiration is where carbon returns to the abiotic environment.

2 Decomposers such as bacteria and fungi; feed on dead organisms and respire, returning carbon to the abiotic environment; they also excrete ammonia from amino acid metabolism; nitrogen-fixing bacteria convert nitrogen into nitrate ions plants can use; to make amino acids/proteins; Rhizobium fixes nitrogen in root nodules of leguminous plants; nitrifying bacteria convert ammonia to nitrites and nitrates; denitrifying bacteria break down nitrates, releasing nitrogen.

3 a Nitrogen fixation by nitrogen-fixing bacteria/ Rhizobium; into ammonia/nitrites/nitrates plants can use; nitrifying bacteria convert ammonia to nitrites and nitrates that most plants prefer; plants use ammonia/nitrites/nitrates to synthesise amino acids and then proteins.

b There is usually more than enough carbon dioxide and water for photosynthesis; in growing seasons the temperature is not a limiting factor; nitrates are needed for protein synthesis; including enzymes for metabolism; proteins for cell growth; protein in seeds/grain/fruit; soil water contains few nitrate ions; which are easily leached/highly soluble; are removed with the harvest.

Deforestation

1 Local extinction of trees, which form the base of food chains and habitats for other species. This reduces the numbers of species and individuals, thus reducing diversity.

2 Fertility depends largely on nitrate ions. These are mainly recycled by decomposition near surface of the soil; often by fungi symbiotic with tree roots. Loss of trees removes decomposing fungi and a source of leaf litter for them to decompose. There is also increased leaching of exposed soils – removing highly soluble nitrate ions.

3 Income from rubber and fruit over 10 years = £2800. Income from clearance and cattle = £670 + £1000 = £1670.

Module 6: Physiology and the Environment B

Transport in plant roots

1 A – endodermis; B – xylem; C – phloem.

2 Water continually moving up the plant; ions dissolved in the xylem; ions actively transported into xylem; by endodermal cells.

3 Water moving through the inter-connecting cytoplasm of cells via plasmodesmata.

4 Casparian strip prevents water moving through the apoplast – forces water to cross cell membrane into the symplast – allowing control of water movement.

Transport in the xylem

1 Measure volume of sap from freshly cut root stump; lower the temperature/deprive the root of oxygen/add cyanide and see if the rate of production slows.

2 Solar energy.

3 Cohesion – attractive force between water molecules due to hydrogen bonding; adhesion – attraction between the water molecules and the xylem walls.

Transpiration

1 Air movements remove water vapour from the leaf surface; increasing the water potential gradient and rate of transpiration. Increase in temperature increases rate of transpiration – provides water molecules with more kinetic energy, allowing them to evaporate more readily.

2 Plants that live in habitats where water is in short supply; having structural adaptations that reduce the rate of transpiration.

3 Thickened waxy cuticle reduces evaporation; curled leaves reduce the surface area for evaporation; reducing the water potential gradient for water loss by, hairs on leaf surface – trap a layer of air which becomes saturated with water vapour, increasing the humidity in air around the stomata – reducing transpiration; reduced leaf surface area over which transpiration can occur; sunken stomata in pits – become saturated with water vapour.

Principles of homeostasis and regulation of blood sugar B 

1 Insulin levels rise when blood sugar above normal – fall when sugar below normal; sugar levels rise after meals – followed by insulin; sugar levels fall when fasting/exercising – followed by insulin.

2 High blood sugar causes pancreas (islets of Langerhans) to secrete insulin into blood; binds to specific membrane receptors on liver cells; causes glucose channels to open; glucose enters liver cells, lowering blood sugar; provides more substrate for enzymes making glycogen.

3 a Liver cells have different specific membrane receptors for insulin and glucagon; these are proteins. Insulin binding causes glucose channels to open. Glucagon binding activates enzymes inside the cells.

b Glucose entering cells gives more substrate for enzymes making glycogen, so more is made. Glucagon activates a phosphorylase that break down glycogen to glucose.

Regulation of body temperature B HB

1 During exercise muscles respire more and produce surplus heat; blood carries heat away; warming of blood detected by thermoreceptors; in hypothalamus; send nerve impulses to heat loss centre in hypothalamus; response coordinated; nerve impulses sent to effectors; responses increase heat loss; circular muscles around arterioles relax → vasodilation; sweat glands → more sweat; muscles at base of hairs relax →hairs lie flat.

2 Cooling blood near stomach causes slight fall in temperature of blood as a whole; this fall detected by receptors in hypothalamus; response is reduction in heat loss from body.

3 Lower body temperature is below optimum for body's enzymes; all enzyme reactions slow, including respiration; less respiration means less heat from respiration; this is what keeps body temperature above environmental; less heat produced than lost, so body temperature continues to fall.

Methods of removing nitrogenous waste

1 Fish can excrete ammonia into the external water, where it is rapidly diluted to harmless levels. They also have few problems of water conservation/loss, since they live in water. Mammals have to conserve water and can not dilute ammonia enough to make it harmless.

2 Insects excrete uric acid which is insoluble in water and non-toxic. It can be stored in the body and then excreted in paste or crystal form; with very little loss of water. This conserves water which is hard to replace.

3 Urea is formed using the amine groups released by the deamination of surplus amino acids in liver cells. This is an alternative to ammonia which is much more toxic than urea. Synthesis is via the ornithine cycle – ammonia reacts with ornithine to make arginine – which is then used in the synthesis of urea and to regenerate ornithine.

Kidney 1 – Urine production

1 Filtrate does not contain blood cells; or blood proteins; does contain dissolved urea, glucose and mineral ions (like plasma); because these substances are small enough to pass through ultrafiltration; in the glomerulus and Bowman's capsule; blood cells and proteins are too large.

2 Presence of glucose indicates too much glucose in filtrate for it all to be reabsorbed; this indicates a very high blood glucose concentration; which could indicate diabetes – a lack of insulin. Protein indicates damage to the glomerulus and Bowman's capsule; the pores have become too large, allowing blood/plasma proteins through; could be due to disease or prolonged high blood pressure.

3 Useful substances reabsorbed along tubule; glucose, Na^+ and K^+ in proximal convoluted tubule; by active uptake; by specific carrier proteins; water reabsorbed by osmosis in; proximal and distal tubule, Loop of Henle and collecting duct; urea not reabsorbed.

Kidney 2 – Water balance

1 Active uptake of chloride ions by cells of loop of Henle; followed by Na^+ ions; gives more negative/lower water potential to surrounding tissues; these also surround collecting duct; water potential lower than filtrate in duct; so water leaves filtrate by osmosis.

2 Filtrate flows in opposite directions in descending and ascending limb – countercurrent; ascending limb impermeable to water but actively transports sodium chloride out; descending limb permeable to water but not sodium chloride; filtrate flowing down descending limb loses water by osmosis; longer the loop, the greater the concentration of sodium chloride that can be achieved; giving very negative water potentials and a greater reabsorption of water by osmosis.

3 Water potential of blood becomes slightly lower/more negative; this stimulus detected by receptors in blood vessels in hypothalamus; response is release of ADH from pituitary gland; ADH makes cells of collecting duct and distal tubule more permeable to water; more water is reabsorbed from filtrate; smaller volume of (more concentrated) filtrate, so less water loss.

Gaseous exchange surfaces

1 Single cells, so large surface area of cell membrane relative to volume of cell; short diffusion pathways between any part of cell and external water.

2 Both have openings in water proof outer covering – stomata and spiracles; opening of these can be controlled to regulate water loss; gases diffuse in gas phase to and from all the cells.

3 Large number of gill filaments – a large surface area; filaments overlap – slowing water flow – allowing more time for gaseous exchange; thin barrier of two cell layers – a short diffusion pathway; gill filaments have many blood capillaries.

Limiting water loss

1 Water from metabolism; water a product of respiration – using respiratory substrate from dry food; long loop of Henle, so able to reabsorb alot of water – produces small volumes of urine.

2 Waxy cuticle over surface of leaves conserves water; stomata for gaseous exchange; on lower surface of leaf, so warm air carrying water vapour is trapped in air spaces in leaf; can reduce size of stomata to reduce loss of water vapour.

3 Cool under stone, so less evaporation loss; inactive under stone, so low respiration, spiracles almost closed and less water loss.

Transport of respiratory gases B HB

1 An increase in CO_2 leads to more carbonic acid production and lower pH; lower pH reduces the affinity of haemoglobin for oxygen; more oxygen released from haemoglobin; dissociation curve shifts to the right/Bohr effect.

2 During exercise the muscles need more energy ; rate of respiration increases; more CO_2 produced, producing Bohr effect.

3 Bright red – haemoglobin; little oxygen in mud at bottom of pond; haemoglobin can bind what oxygen there is.

4 Fetal haemoglobin has a higher affinity for oxygen than mother's haemoglobin; so at same concentration of oxygen, will have higher saturation; so always diffusion/ concentration gradient for oxygen from mother to fetus.

Digestion and absorption of food B HB

1 Salivary amylase is inactivated/denatured in the stomach, so need a second amylase to complete hydrolysis of starch to maltose; two enzymes acting on starch makes sure all is converted to maltose; may be different types of starch, with a different amylase for each.

2 Endopeptidases; hydrolyse bonds in the middle of proteins, producing more 'ends' for exopeptidases; different endopeptidases cut in different places, so more 'ends', e.g. pepsin in stomach/trypsin in small intestine; exopeptidases remove short peptides from 'ends' of polypeptides, giving lots of substrate for peptidases inside epithelial cells of small intestine.

3 Fatty acids and glycerol associate with bile salts and cholesterol to form micelles; these lipids diffuse into epithelial cells across cell membrane; re-synthesised into triglycerides and phospholipid; form chylomicrons, secreted into lacteals.

4 Antibiotics may kill bacteria in the rumen; so no cellulase production; so cattle unable to properly digest their food.

Control of digestive secretions B HB

1 Receptors in retina/eye send nerve impulses to the brain; learning leads to association area linking stimulus to food; sympathetic nervous system; nerve impulses along vagus nerve; to salivary glands/stomach/gallbladder.

2 Stomach stretches as food enters, receptors detect this and leads to increased peristalsis in ileum, to move food along and make room for food from stomach; chyme leaving stomach stretches duodenum, leading to reduced release of chyme from stomach.

3 Fat-poor food causes secretion of secretin; causes release of pancreatic juice with hydrogencarbonate ions to neutralise acid from stomach. Fat-rich causes secretion secretin and cholecystokinin-pancreozymin; causes release of pancreatic juice with lipase (amylase and trypsin); and makes gallbladder contract, releasing bile.

Histology of the ileum  

1 Large surface area relative to the volume of the body; produced by being long, having folds on inner surface, villi on folds and microvilli on epithelial cells; short diffusion pathways across the epithelium and to the capillaries; good blood supply, capillary networks in each villus; flow of blood removes absorbed products of digestion and maintains steep diffusion gradients; epithelial cells' membranes have channel and carrier proteins.

2 Some cells, secrete mucus to protect the gut lining; produce intestinal juice with enzymes; have channel/carrier proteins in their cell membranes; have digestive enzymes/maltase in their cell membrane; have microvilli, to increase surface area for absorption; have lots of ER/Golgi, for secretion of protein/mucus; lots of mitochondria, to supply energy for synthesis/active transport.

Metamorphosis and insect diet

1 Pupa.

2 Larva is growing, so needs all types of biological molecule; so need range of enzymes to obtain these by hydrolysis of all major food molecules. Adults are not growing, just looking for a mate; so need sugar for respiration, to provide energy for flying; only need enzyme to digest sucrose from nectar.

Neurones and action potentials B HB

1 Long axon, carries nerve impulses long distances; axon myelinated – nerve impulses 'jump' from one node of Ranvier to the next $\rightarrow$ fast transmission; many dendrites, to synapse with other neurones.

2 Travelling depolarisation; transmembrane protein channels open for Na^+; they diffuse in along their diffusion gradient; resting potential across membrane falls locally – depolarisation; inrush of sodium ions makes potassium channels open; they diffuse out along diffusion gradient; start to restore membrane potential; sodium channels close and sodium pump pumps out sodium ions again; the local depolarisation opens sodium channels in next section of axon, so depolarisation moves along.

3 Refractory period 3 milliseconds; so 333 nerve impulses per second.

4 Small stimuli below threshold for initial depolarisation of cell membrane, so no action potentials; information on size of stimulus carried as the frequency of nerve impulses; larger the stimulus, greater the frequency of nerve impulses; refractory period means limit to frequency of nerve impulses and size of stimulus we can detect.

Synoptic transmission  

1 Action potential reaches presynaptic membrane; depolarisation opens calcium channels – they diffuse in; vesicles with acetylcholine fuse with membrane; acetylcholine released and diffuses across narrow synaptic cleft; binds to specific receptors on postsynaptic membrane; makes Na^+ channels open; leads to depolarisation and if enough of these, an action potential.

2 Information carried across synapse by bursts of acetylcholine, corresponding to action potentials at presynatic membrane; this produces distinct depolarisations in postsynaptic membrane; if transmitter remained in synaptic cleft, constant depolarisations and no information.

3 Excitatory produce epsp and inhibitory prevent these; depends on frequency of each; and on spatial summation; and temporal summation.

4 Venom molecule has part with a shape that fits acetylcholine receptor; competes with acetylcholine and occupies binding sites; does not cause depolarisation of postsynaptic membrane; but stops acetylcholine, so no action potentials and paralysis.

Receptors B HB

1 Dermis of skin; joints and tendons.

2 Pressure changes shape of sodium channels; they open, leading to generator potential; if enough pressure, enough channels open to produce an action potential; theshold.

The retina B HB

1 Accommodation/fine-focusing; ciliary muscles contract; taking tension out of suspensory ligaments; elasticity of lens makes it more biconvex/fatter; shortens focal length of lens/keeps image of print in focus on retina.

2 Light bleaches the pigment rhodopsin in membranes of vesicles; rhodopsin breaks down to retinene and scotopsin; alters permeability of membrane to Na^+ ions; nerve impulses formed that pass along optic nerve; carrying information.

Monochromatic and colour vision B HB

1 As in the graph on page 76 but green and red-sensitive lines on top of each other.

2 Little light at night, so only rods stimulated; outside of the fovea, in edge of vision; when looking straight at, light falls on fovea; only cones there and they do not react to dim light.

3 Light focussed on fovea where cones are; cones give colour vision; each cone synapses with one bipolar and one ganglion cell; so sends information along optic nerve to brain about its very small area of the retina; gives high visual acuity.

Autonomic nervous system B 

1 This is a simple reflex; involves three neurones and the spinal cord; sensory, relay and motor neurones; brain informed from relay neurone; by which time response already taking place.

2 Dilation of pupil, to gather more visual information; dilation of bronchi, to get more oxygen to lungs – for respiration; heart rate increases, increasing oxygen and glucose to muscles for respiration – providing energy for contraction; vasoconstriction of blood vessels, to increase blood pressure and thus rate of flow to muscles; also reduces blood flow to skin – in case of cuts; diverts blood from skin to muscles; reduces blood flow to gut, so more blood to flow to muscles – so more oxygen and glucose for respiration.

3 Contracting muscles press on veins; forces blood back towards the heart; fills ventricles more; making them beat harder and faster.

Behaviour B HB

1 Negative phototaxis; make them move into tunnel/ground; away from predators.

2 Light, because they might move away from it/in response to it. Number of woodlice, because they might clump together/avoid each other. Another suitable environmental factor, with explanation.

Module 7: The Human Lifespan HB

Gamete formation and fertilisation

1 Mitosis of spermatogonia; growth of some into primary spermatocytes; meiosis to form secondary spermatocytes, then spermatids; maturation to form spermatozoa.

2 Any from the table on page 86.

3 Contact between oocyte and sperm causes acrosome reaction; digestive enzymes released, to digest protective layers round oocyte; so membrane of sperm and egg oocyte can touch and fuse.

Implantation and the fetus

1 Implantation is when the blastocyst (embryo) attaches itself to the endometrium; blastocyst secretes digestive enzymes, to digest its way into the endometrium.

2 Placenta has chorionic villi and microvilli, to give large surface area for exchange; few cell layers between maternal and fetal blood, giving a short diffusion pathway; many mitochondria, to provide energy/ATP for active transport; maternal and fetal blood flow in opposite directions/counter-current, maintaining steep diffusion gradients.

3 Fetal blood entering the right ventricle comes from the placenta, so contains lots of oxygen; this blood pumped out along pulmonary artery, but most is diverted into the aorta by the ductus arteriosus; because the lungs are not working as for gaseous exchange and oxygenated blood in aorta gets to the rest of the body.

Changes during pregnancy

1 Greater rate of blood flow to the placenta; this increases delivery of oxygen and nutrients and removal of carbon dioxide and waste products; which increases exchange rates across the placenta to and from the fetus.

2 Progesterone: (**a**) inhibits release of FSH from the pituitary gland; this prevents the development of any primary follicles and eggs in the ovaries, (**b**) maintains the endometrium.

3 Prolactin stimulates production and release of milk by mammary glands. Oxytocin stimulates milk ejection/release from the nipple.

Growth and development

1 Measure a particular parameter at regular time intervals during the growth of individuals of a population.

2 Cross-sectional studies involve measurements on large samples of individuals of certain ages in a population; all the measurements can be done in a short time, so the results are quickly available; but it assumes that the samples of people of different ages come from the same genetic and environmental backgrounds – which may not be true.

3 Different organs grow at different times and rates during the development of an individual; this will change the body proportions of the organs over time.

Growth and puberty

1 Growth hormone causes the liver to produce somatomedins; and thus stimulates growth and division of bone-producing cells in the ends of long bones; so its lack will reduce bone growth.

2 Thyroxine stimulates the rate of respiration in cells; too much will cause more respiration than required and use lots of respiratory substrates; these come from food and the body's reserves, causing weight loss.

3 In females, FSH stimulates oestrogen secretion from ovaries, which then stimulates secondary sexual characteristics; in males, LH stimulates testosterone secretion by testes, which stimulates secondary sexual characteristics.

Principal nutrients in the diet

1 Essential amino acids can not be made in the body, they have to be present in the diet.

2 Vitamin D is needed for absorption of calcium and phosphate ions from the gut, which are needed for strong bones and teeth. Iron is needed for the production of haemoglobin and myoglobin.

3 Sample in test tube with ethanol. Shake to dissolve any fat. Add water and shake, to see if a white emulsion forms.

Dietary requirements

1 This is the metabolic rate at rest.

2 They may not contain: correct proportions of the different food types, causing ill-health; enough energy, so muscle and tissue protein is used for respiration – causing wasting. May also cause: vitamin and mineral defficiencies; increased infection rate; low blood pressure; constipation; halt to menstruation.

3 Glycogen can be hydrolysed to release glucose. This can be used to maintain a high rate of respiration.

Muscle structure and the sliding-filament theory of muscle contraction

1 **a** A-band – no change; **b** H zone reduced.

Control of muscle contraction and muscles as effectors

1 Nerve impulses at neuromuscular junction cause calcium ions to enter muscle fibres; calcium ions bind to troponin, which is attached to tropomyosin; binding causes 3D shape changes, so tropomyosin moves; off myosin binding site on actin, which allows myosin head to bind – producing contraction.

2 ATP from respiration; myosin head has ATPase activity; uses ATP to detach from actin; phosphocreatine can be used to phosphorylate ADP to ATP.

3 Slow fibres tire slowly, because they: respire aerobically; have myoglobin to store oxygen.

Senescence

1 Cardiac muscle fibres and output decrease; speed of nerve impulse conduction and reflexes slow; basal metabolic rate drops.

2 The menstrual cycle slows, then stops; the number of viable primary follicles in the ovaries decreases; so oocyte development stops and ovulation.

Answers to end-of-module questions

Module 5: Inheritance, Evolution and Ecosystems B HB

1 a i Haploid number = 22.
ii Diploid number = 44.

b Homologous chromosomes have the same genes at the same loci; but not necessarily the same alleles of the genes.

c Chiasmata form, involving tangling of non-sister chromatids from the homologous cromosomes; crossing over may take place, where parts of non-sister chromatids are exchanged.

d Sexual reproduction involves the fusion of gametes at fertilisation; these contain the haploid number of chromosomes, so fusion produces the diploid number; meiosis occurs at some stage, to ensure that gametes contain the haploid number of chromosomes.

2 a i One form of a gene.
ii I^0, because genotype I^AI^0 is type A and I^BI^0 is type B.
iii I^A and I^B are codominant, because I^AI^B is type AB.
iv Discontinuous variation; it is qualitative; there are discrete classes with no intermediates.

b Both parents must carry I^0. So, parental genotypes must be I^AI^0 and I^BI^0.

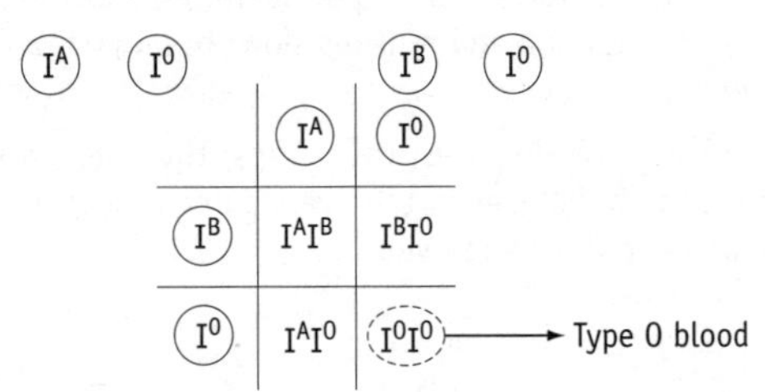

3 a Random/chance; mutation.

b Let p = the probability of the dominant allele W. Let q = the probability of the recessive resistance allele w. Probability of homozygous ww = q x q , or q^2. Percentage of homozygous rats is 1%, which is a probability of 0.01. Therefore, the probability of q = $\sqrt{0.01}$ = 10% and probability of p = 90%. Probability of heterozygous is 2pq = 0.18%.

c Rats with resistance allele survived to reproduce; they produce the new population; size not as large as before, because resistant rats not as well adapted as normal rats to other environmental factors.

4 a i A group of organisms of the same species, living in the same habitat and able to interbreed.
ii Where an organism is found in the habitat and its role in the community – includes the range of abiotic factors it needs and feeding requirements.

b Frame quadrats; set up grid in each area and give numbers to intersections; use random number generator/tables to select intersects; put quadrat on intersect and record percentage cover for each species; use large number of quadrats (20+); compare average percentage cover for species.

c i Number of each species and number of species.
ii Desert a much more severe environment; fewer species with adaptations to survive; so fewer species in desert; also fewer number of each species likely in desert; so diversity lower in desert.

5 a i Reduced NADP; ATP; oxygen.
ii In the grana of the chloroplast.

b i 5 carbons
ii 3 carbons.

c Reduced NADP, used to reduce GP to carbohydrate.

Module 6: Physiology and the Environment B

1 a W – epidermis; X – endodermis; Y – phloem; Z – xylem.

b Water uptake mainly by root hairs; by osmosis; along water potential gradient; soil water higher water potential than root hairs; due to active uptake of mineral ions by hair cells; water moves along water potential gradient across cortex to the xylem vessels; via apoplast, symplast and vacuolar; enters symplast in endodermal cells; because of Casparian strips; water potential gradient is maintained by water continually moving up the xylem and by dissolved ions in the xylem sap.

2 a Cohesion-tension hypothesis; heat energy evaporates water from leaves; from mesophyll cells next to air spaces; water vapour diffuse out through the stomata into the air; transpiration; water potential of mesophyll cells lower than inner mesophyll cells; water moves by osmosis; along water potential gradient; by apoplast, symplast and vacuolar pathways; drawing water from xylem; creating a tension; 'pulling up' water and dissolved ions; continuous water column from leaves to roots; maintained by cohesive forces between water molecules; adhesive forces between water molecules and xylem walls; due to hydrogen bonding.

b The following increase the rate: light – opens stomata; allowing more water vapour to diffuse out: temperature – increases kinetic energy of water molecules; increasing rate of evaporation and diffusion of water from stomata: humidity – increase decreases water potential gradient; so slower rate of diffusion from stomata: air movement – wind removes water vapour from stomata; which increases water potential gradient and diffusion.

c Thick waxy cuticle, to reduce cuticular evaporation loss; hairs on surface, to trap still layer of air – saturates with water vapour and reduces loss; curled leaves, to reduce surface area/increase water vapour round stomata; reduced leaf surface area, to reduce area for evaporation; sunken stomata, to trap still layer of air – saturates with water vapour and reduces loss.

3 a i insulin binds to receptor; which opens glucose carrier, so glucose enters liver cell; increases substrate for enzyme making glycogen;

ii Glucagon binds to receptor; causes activation of enzyme in liver cell; converting glycogen to glucose, which diffuses out into blood.

b Build-up of glycogen in cells; low blood sugar concentration; weakness and fatigue.

4 a Leaf type X; greater surface area for stomata/gas exchange; same volume as Y; so X larger surface to volume ratio.

b Type Y, because thicker cuticle – to reduce cuticular evaporation; smaller surface area to volume ratio; so less surface area for stomata.

5 a Y is ammonia; Z is uric acid;

b Fish.

c Endothelial cells of glomerulus have pores; basement membrane of capillaries act as filter; podocytes of Bowman's capsule have small gaps between them; all these act as a filter, allowing blood plasma through – except for blood cells and large blood proteins; high blood pressure in glomerulus forces liquid through filter – ultrafiltration.

d i From metabolism; mainly from respiration.

ii Losses from faeces, lungs and skin = 0.043 + 73.2 = 73.24%. Therefore loss from urine = 100 –73.24 = 26.76% of 60cm^3. This equals 16.01 cm^3.

6 a Stomach – pepsin; endopeptidase that hydrolyses; bonds in polypeptides to produce shorter polypeptides; continued by endopeptidase trypsin in small intestine; endopeptidases speed up digestion by making more ends for exopeptidases to act on; by removing short peptides and amino acids; which are then digested b peptidases in epithelial cells.

b Fat-rich chyme enters small intestine; certain cells of small intestine secrete cholecystokinin-pancreozymin; into the blood; causes release of pancreatic juice.

7 a At A, sodium channels open; sodium ions diffuse in along concentration gradient; causing depolarisation of membrane.

b Potassium ion channels open; potassium diffuse in and starts re-polarisation; sodium channels close; sodium 'pumps' start to actively transport sodium ions out across the membrane.

c 'About' 320 per second; 1000 milliseconds divided by the refractory period; about 3.2 milliseconds.

8 a Sympathetic dilates pupil – increases information from light receptors – parasympathetic makes pupils contract, to protect the receptors in the retina from too much light energy; sympathetic dilates bronchi, so more/faster delivery of oxygen for respiration – parasympathetic constricts – to reduce water loss; sympathetic causes vasoconstriction, diverting blood to muscles/ reducing blood loss from cuts on skin – parasympathetic causes vasodilation, so more blood supplied to gut for digestion.

b During exercise muscles contract more, and so press on veins more; the extra pressure forces blood towards the heart; this fills the ventricles more; which makes them beat harder and faster.

Module 7: The Human Lifespan HB

1 a i Insulin binds to receptor; which opens glucose carrier, so glucose enters liver cell; increases substrate for enzyme making glycogen.

ii Glucagon binds to receptor; causes activation of enzyme in liver cell; converting glycogen to glucose, which diffuses out into blood. b Build-up of glycogen in cells; low blood sugar concentration; weakness and fatigue.

2 a Stomach – pepsin; endopeptidase that hydrolyses; bonds in polypeptides to produce shorter polypeptides; continued by endopeptidase trypsin in small intestine; endopeptidases speed up digestion by making more ends for exopeptidases to act on; by removing short peptides and amino acids; which are then digested by peptidases in epithelial cells.

b Fat-rich chyme enters small intestine; certain cells of small intestine secrete cholecystokinin-pancreozymin; into the blood; causes release of pancreatic juice.

3 a Sympathetic dilates pupil – increases information from light receptors – parasympathetic makes pupils contract, to protect the receptors in the retina from too much light energy; sympathetic dilates bronchi, so more/faster delivery of oxygen for respiration – parasympathetic constricts – to reduce water loss; sympathetic causes vasoconstriction, diverting blood to muscles/reducing blood loss from cuts on skin – parasympathetic causes vasodilation, so more blood supplied to gut for digestion.

b During exercise muscles contract more, and so press on veins more; the extra pressure forces blood towards the heart; this fills the ventricles more; which makes them beat harder and faster.

4 a i Calcium ions cause vesicles to move to presynaptic membrane; membrane of vesicle fuses with presynaptic membrane, releasing acetylcholine.

ii Diffuses rapidly across synaptic cleft; binds to specific receptors on postsynaptic membrane; broken down by acetylcholine esterase.

b i Break down of synapsin means vesicles can not fuse with presynaptic membrane; acetylcholine not released; no transmission across synapse/paralysis/ coma.

ii Endopeptidase hydrolyses bonds between amino acids; at points along length of protein, to produce shorter polypeptides.

c Synapses with more than one other neurone on a neurone; action potentials/nerve impulses arriving at these synapses lead to additive effect to produce greater depolarisation; leading to threshold reached for action potential.

5 a Sodium ion channel proteins open; sodium ions diffuse in down their concentration gradient; the membrane depolarises.

b Potassium channel proteins open; potassium ions diffuse out across the membrane down their concentration gradient; starting re-polarisation of the membrane; sodium channels close.

c Refractory period 2.2 milliseconds, so maximum frequency is 1000 ÷ 2.2 = 455 nerve impulses per second.

6 a i Birth 2000 kJ per day, three years 6000 kJ per day – so % increase is 6000 – 2000 ÷ 2000 x 100 = 200%.

ii Up to age 7 is before puberty; and then the affect of hormones is different in men and women; age 15 and onwards are post puberty.

b i Fairly active, due mainly to higher BMR of men – who are on average larger.

ii Very active people have higher energy demands for muscle contraction/have more muscle; this is true for both sexes/ages.

c Over 65 year-olds have lower BMR; have lost some muscle.

7 a Sliding-filament hypothesis; head of myosin interacts with actin filament to form actomyosin bridge; head of myosin moves; actin filament is pulled past myosin; ATP used to release myosin from actin; so that it can react again with actin; idea of ratchet mechanism/ actin moves past myosin in one direction; actin filaments attached to Z lines either end of sarcomere; actin each side pulled over myosin filaments in centre of sarcomere, so sarcomere shortens.

b i X is fast muscle fibre, Y is slow.

ii M has fast and slow fibres; fast for running; slow for posture and walking depending on how fast you walk.

8 a Two of: mitosis before birth in woman and from puberty to old age in men; meiosis continually in man, 1st division before birth in woman and 2nd at fertilisation; in man meiosis produces 4 spermatids/sperm, in woman produces 1 egg and 2 polar bodies; one egg produced per month, sperm production continuous.

b i Foramen ovale, allows blood to flow from right to left atrium; this is oxygenated blood from the placenta; ductus arteriosus allows blood to flow from the pulmonary artery to the aorta; both divert a lot of oxygenated blood away from the lungs which are not being used for gaseous exchange; and sends it directly to the rest of the body via the aorta.

ii It would reduce the blood supply to the lungs; this would reduce ability to take up oxygen for respiration; this would lead to varying degrees of weakness and tiredness.

Index

A
abiotic factors 20
  and adaptation and diversity 21, 22-23
  and deforestation 36
  interactions with biotic factors 34-35
  and succession 24-25
  and transpiration rate 46
absorption
  of food 40, 64-65, 68
  by kidneys 55, 56-57
  by plant roots 43
accommodation of the eye 75
acetylcholine 72, 73, 78
acetylcoenzyme A 32, 33
acrosomes 86, 87
actin 84, 100, 101, 102
action potentials 41, 70-71, 72-73
activation energy 27, 30
active transport
  across cell membranes 68, 70-71, 89
  across placenta 88-90
  of chloride ions in kidney 56-57
  in the distal convoluted tubule 55
  of mineral ions in plants 43, 44
adaptation
  and diversity 22
  for gas exchange 58
  of haemoglobin for low oxygen 63
  of ileum 68
  and natural selection 15, 16-17
  physiological 22
  of placenta 88-90
  and variation 10
  for water shortages 40, 47, 60, 61
adipose tissue 96
adolescents 93, 94, 98
ADP 30, 33
adrenaline and adrenal glands 72, 79
adults, human 93, 98, 99
aerobic respiration see respiration
aerotaxes 80
ageing (senescence) 84, 98-99, 104
agonists 73
air movement, and transpiration rate 46
alanine 96
albinos, probability of 14
alleles 4, 5
  and inheritance 6-7, 8-9, 14
  recessive 4, 6, 7, 14
  for sickle-cell anaemia 17
  see also genes; genotypes; mutation
allometric growth 93
amino acids
  changes during mutation 12-13
  digestion in mammals 53
  and nitrogen cycle 34, 35
  as nutrients 96
  as respiratory substrates 31, 32
  transport across placenta 89
  as waste products 52, 53
ammonia 34-35, 52, 53, 65
amoebae (*singular* amoeba) 58
amylase 64, 67
anaemia 97
anaerobic respiration 31, 32, 103
animals
  body temperature regulation 50-51
  classification 19
  desert-living 57, 60
  interspecific competition 22
  respiratory quotients 31
  trapping methods for 21
  water loss 60
  see also mammals
antagonists 73
antibiotic resistance 16, 17
antibody transport across placenta 89
apoplast pathway of water transport 43, 44
ATP 30
  from photosynthesis 2, 26, 27
  in muscle contraction 101, 102, 103
  in respiration 3, 30, 32-33
ATPase 101, 102, 103
autonomic nervous system 41, 78-79
  and digestion 66
  neurotransmitters in 72, 78
auxotrophs 28
axons 70-71

B
bacteria 34, 35, 65, 80
banding in skeletal muscle 100, 101
basal metabolic rate 50, 51, 84, 94, 98, 104
behaviour
  control 41
  innate (genetically determined) 80-81
  regulating body temperature 50, 51
Benedict's test for glucose 97
bile and bile salts 65, 66
binomial nomenclature of species 18
biotic factors 20, 22, 36, 34-35
birds 52
birth, hormonal changes at 91
biuret test for proteins 97
bivalent chromosomes 5
blastocysts 84, 88, 91
blood
  circulation
    autonomic control 79
    countercurrent system in fish 59
    fetal/maternal 63, 84, 88-89
    for gas exchange in large animals 58
    in the kidney 54
    for respiratory gas transport 62-63
    for waste product transport 53
  fetal 63, 88
  glucose content regulation 48-49
  groups 7, 10, 12
  homeostasis 40, 48-49
  plasma 54, 55, 56, 62, 90
  in pregnancy 90
  pressure 79
  vessels, constriction/dilation 50, 51, 79
Bohr effect 62
Bowman's (renal) capsule 54
bronchi, autonomic control 79
butterfly, life cycle 69

C
calcium ions 72, 97, 99, 102
calorimeter/calorimetry 98
Calvin cycle 27
capacitation in spermatogenesis 85
capillarity in plants 44
capture–recapture techniques 21
carbohydrates
  digestion and absorption 64
  as nutrients 96, 99
  as respiratory substrates 31, 32, 96, 98
carbon cycle 3, 34, 37
carbon dioxide
  from deforestation 37
  from respiration 33, 34
  in photosynthesis 27, 34
  transport in the body 40, 62, 89
carbonic anhydrase 62
cardiac centre 79
  see also heart
carnivores 53
carrier proteins 48, 55, 68, 70-71, 89
Casparian strip 42, 43
cation pumps 70, 71
cell membranes 68, 87, 89, 94, 96
  transmission of nerve impulses 70-71, 72-73
cellulose 43, 44, 65
centromeres 5
chemical energy
  from photosynthesis 26, 28
  from respiration 30
chemotaxes 80
chewing 64
chi-squared test 8-9
chiasmata 5, 13
children 92-93, 94, 98-99
chloride ions 56-57
chlorophyll 26, 27, 30
chloroplasts 27, 46
choice chambers 80
cholecystokinin-pancreozymin 67
cholesterol 96
cholinergic synapses 72-73
chorionic gonadotrophin (HCG) 88, 91
chromatids 5
chromosomes 4-5, 13
chylomicrons 65
chyme 65, 66, 67
ciliary muscles 75
classification, biological 2, 18-19
climatic factors, and diversity 22
climax communities 2, 24-25, 36
codominant alleles 6-7
coenzyme A 32, 33
coenzymes 26, 32-33
  reduced 2, 3, 26, 32-33
cohesion tension hypothesis 44-45
collecting duct 54, 55, 56
colonisers 24
colour vision 76-77
communities 20-21
  climax 2, 24, 25
  diversity of 22-23
  succession of 24-25
companion cells in phloem 42
competition 10, 22, 24-25
conditioned reflexes 66
cone cells 41, 75-76, 77
continuous variation in populations 10, 12, 92
convoluted tubules 54, 55, 56
cornea 75
corpus luteum 91, 95
countercurrent circulation
  blood/water system in fish 59
  fetal/maternal blood in humans 88
countercurrent multiplier hypothesis 57
cretinism 94
cristae (*singular* crista) 33
cross bridges in skeletal muscle 101, 102
cross-sectional studies 92-93
crossing over in meiosis 5, 13, 86
crypts of Lieberkühn 68
cystic fibrosis 6
cytoplasm, glycolysis in 32

D
Darwin's finches 15
deamination 35, 53
decomposers/decomposition 28, 34-35
deforestation 3, 36-37
denitrifying bacteria 34, 35
depolarisation
  and action potential 70-71, 73
  in Pacinian corpuscles 74
  of sinoatrial node 79
development 92
  at puberty 95, 99
  of embryos 84, 87
  of fetuses 84, 88
  of gametes 84-86

diabetes 48, 49
diet
  and blood glucose levels 48
  of carnivores 53
  and energy requirements 84, 98-99
  of insects 41, 69
  principal nutrients in 96-97
  see also food
diffusion
  across placenta 88-89
  in cell membranes 70-71, 72-73
  in ileum 68
  in insects 61
  in leaves 46, 47, 58, 61
  rate 58, 61, 71
digestion 40, 53, 64-65, 66
  in butterfly larvae 69
  control of 66, 67
digestive enzymes 34, 40, 64-65, 67, 68, 69
  of bacteria and fungi 34
  of butterfly larvae and adults 69
  in humans 40, 64-65, 67, 68
  in sperm cells 86, 87
dihybrid crosses 8-9
diploid organisms 4, 6, 14
directional selection 16-17
directional stimulus 80
discontinuous variation in populations 10, 12
disruptive selection 17
distal tubule 54, 55, 56
diversity 2, 22-23
  reduction by deforestation 36, 37
DNA 4, 5, 12-13, 99
dominant alleles 4, 6, 7, 8
ductus arteriosus 89
duodenum 67

E
ecological techniques 20-21
ecosystems 2, 20-21
  and deforestation 36-37
  diversity in 22-23, 36
ectotherms 50
egg formation (oogenesis) 87
electron carrier/transport chains
  in photosynthesis 26-27
  in respiration 33, 60
embryos 84, 87
emigration 15
emulsion test for lipids 97
endodermis in plant roots 42, 43
endometrium 91
endopeptidases 64, 65
endotherms 50-51
energy
  measurement by calorimetry 98
  pyramid of 29
  requirements and diet 84, 98-99
  transfer
    in ecosystems 3, 20, 28-29
    efficiency 28
    in photosynthesis 26, 28
    in respiration 33
environment
  abiotic *see* abiotic factors
  adaptation to 15, 22
  and deforestation 36-7
  and diversity 2, 22-23
  and natural selection 15, 16-17
  and survival rates 16-17
  and transpiration rate 46
  and variation in populations 10, 12, 13
enzymes
  converting glucose to glycogen 48
  digestive *see* digestive enzymes
  in photosynthesis 26, 27
  rate of reaction 24, 34, 37, 49, 50
  and temperature 24, 34, 37, 48, 50
epididymis 85
epithelial cells/epithelium (*plural* epithelia)
  and digestion 64, 65, 68
  and gas diffusion 58
  germinal 87
eukaryotic cells 19

evaporation
  from plants *see* transpiration
  of sweat 51
evolution 14, 18, 19
excitatory synapses 72-73
excretion of waste products 28, 35, 52-53
exercise
  Bohr effect 62
  and dietary requirements 99
  and heart rate 79
  and respiration 51, 61
exopeptidases 64, 65
eye 75, 76-77, 79

F
F_1 and F_2 hybrids 6, 8
fallopian tubes (oviducts) 87
families 18-19
farming
  and deforestation 36
  and natural selection 17
fats *see* fatty acids; lipids
fatty acids
  from digestion of lipids 65
  as respiratory substrates 31, 96
  transport across placenta 89
feedback
  negative 48-49, 51, 67, 95
  positive 67, 91
fertilisation 4, 13, 87
fetuses
  development 84, 88-90
  fetal/maternal blood circulation 63, 84, 88-90
  haemoglobin 63, 88
Fick's law 58
fish 52, 59, 63
flagella (*singular* flagellum) 86
focusing with the eye 75
Follicle Stimulating Hormone (FSH) 91, 95, 104
follicles, ovarian 87, 91, 95
food
  digestion 40, 53, 64-67
  nutrient cycles 34-35
  tests 97
  see also diet
food chains and webs 28
  and deforestation 36
  and diversity 22, 23
  and energy transfer 20, 28-29
  and natural selection 16
foramen ovale 89
fovea 75, 76
frame quadrats 20-21
fungi 19, 34, 35, 37

G
gallbladder 65, 66, 67
gametes 4, 84, 85
  formation (gametogenesis) 85-87
  random fusion 13
  representation in Punnett squares 8
ganglion cells 76, 77
gas exchange surfaces 40, 58-59
gastric juice 64, 66, 67
gastrin 67
genera (*singular* genus) 18-19
generator potential 74
generic names 18
genes 4
  antibiotic-resistant 16-17
  in butterfly larvae and adults 69
  formed by mutation 12-13
  gene pools 2, 14, 37
  genetic code 12, 14
  number of 4-5
  see also alleles; genotypes
genotypes 4
  of fetus and mother 89
  and Mendelian inheritance 6, 7
  of sperm cells 86
  variation 12-13
  see also alleles; genes
geographical isolation 15
gills 52, 59

glomeruli (*singular* glomerulus) 54
glucagon 48-49
glucose 31, 96
  Benedict's test for 97
  control in blood 40, 48-49
  as digestion product 64
  from photosynthesis 27, 30
  in respiration 30, 31
  supply to fetus 88, 89
  test for 97
glycerate 3-phosphate (GP) 27
glycerol 65, 89
glycogen 96, 99, 103
glycolysis 32
glycoproteins 96
Golgi bodies 68
gonadotrophins 88, 91, 95, 104
granum membranes 26, 27
grasses, succession of 24, 25
growth
  absolute 92
  at puberty 94-95
  and dietary requirements 99
  in gametogenesis 85, 86
  hormone (PTH) 94, 95
  human variation in 12
  in life cycle of butterflies 69
  limitation of 22, 34
  of plants 22, 24
  rates and measurement 84, 92-93

H
habitats 20
haemoglobin
  fetal 63, 88, 99
  and iron requirement in diet 99
  oxygen transport by 40, 62-63, 88
  in sickle cell anaemia 13, 17
haemophilia 7
hair, and body temperature regulation 51
hand withdrawal reflex 78
haploid cells 4-5
Hardy–Weinberg principle 14
heart
  autonomic control 79
  cardiac output 90, 104
  changes in pregnancy 90
  fetal 89
heat
  loss 28, 33, 50-51
  measurement by calorimetry 98
  receptors 50, 78
  transfer 28
height
  measurement 92-93
  variation 10, 12, 92
herbivores 65
heterozygous organisms 4, 6, 14
hierarchies 18
homeostasis 40, 48-49
homologous chromosomes 5, 13
homozygous organisms 4, 6, 14
hormones
  and blood glucose control 48-49
  and body temperature control 51
  and digestion control 66, 67
  and growth at puberty 94-95
  in pregnancy 90-91
  secreted by placenta 88, 91
human chorionic gonadotrophin (HCG) 88, 91
humans
  classification 18
  growth 10, 12, 16, 92-93
  respiratory quotients (RQ) 31, 98
  sexual reproduction 84, 85-87
humidity, and transpiration rate 46
humus 24
Huntington's chorea 4
hydrochloric acid, and digestion 64, 66, 67
hydrogen bonds in water 44, 45
hydrogencarbonate 62, 65, 67
hydrolysis 40, 64-65, 101, 102
hyperpolarisation of cell membranes 73
hypothalamus 50, 78

I
ileum 66, 68
imago (adult butterfly) 69
immigration 15
immune response 86, 89, 93
implantation of blastocysts 88
independent assortment 5, 8, 13, 86
inheritance
  of behaviour 80
  genetic (Mendelian) 2, 6-7, 8-9
inhibitory synapses 73
innate behaviour 80-81
inorganic ions 22, 34, 97
  see also mineral ions
insects 41, 52, 59, 61, 69
insulin 48-49
interspecific competition 22, 24-25
intraspecific competition 10, 22
iodine test for starch 97
ion channels 70-71, 72, 73, 74
iron, as a nutrient 97, 99

K
kidney function 40, 54-57
kineses (*singular* kinesis) 41, 80
kingdoms 2, 18-19
Krebs cycle 32-33
Kwashiorkor 99

L
lactation 91, 99
lacteals 65, 68
larva (caterpillar) 69
leaves
  adaptation to water shortage 47, 61
  gas exchange in 58-59
  water loss from 46, 47
  water transport in 43, 44-45, 61
Lepidoptera 69
light
  and diversity 23
  and photosynthesis 2, 26-27, 30
  and phototaxis 80
  receptors 41, 75
  and transpiration rate 46
Lincoln index 21
line transects 20-21
lipase 65, 67
lipids
  digestion 65
  emulsion test for 97
  as nutrients 96
  as respiratory substrates 31, 32, 96, 98
  transport across placenta 89
liver function 40, 48-49
llamas 63
longitudinal studies 92
Loop of Henle 54, 55, 56-57
lungs, fetal 89
Luteinising Hormone (LH) 95, 104
lymphoid tissue 93

M
maggots 80
malaria resistance 17
maltase 64
mammals 50-51, 52-53, 63
  see also animals
mammary glands 90, 91
mark–release–recapture techniques 21
marker genes 17
marram grass 24
maturation, in gametogenesis 85, 86
medulla 78, 79
meiosis 2, 5-6, 8, 13, 85, 86, 87
Mendelian (genetic) inheritance 2, 6-7, 8-9
menopause 84, 104
menstruation 91, 99, 104
mesophyll cells 43, 44, 46
mesophytes 47, 61
metabolic rate, basal 50, 51, 84, 94, 98, 104
metamorphosis of insects 41, 69
microvilli 68, 88
mineral ions 24, 42-43, 44-45, 56-57, 89
  see also inorganic ions
mitochondria (*singular* mitochondrion)
  and digestion 68
  in placental epithelial cells 88
  and respiration 32, 33
  in skeletal muscle 100, 101, 103
mitosis 85, 86, 87, 88
monohybrid crosses 6
moths 80
motor neurones 70, 78
multiple alleles 6-7
muscles
  ciliary 75
  contraction 101-103
  respiratory substrates of 31
  skeletal 84, 100-103
mutation 12-13, 15, 16-17
myelin sheath of neurones 70, 96
myofibrils 100
myoglobin 103
myosin 84, 100, 101, 102

N
NADP and NADPH 26, 27
natural selection 15, 16-17, 80
negative feedback 48-49, 51, 67, 95
nephrons 54
nerve impulses 41
  and ageing 104
  conduction along axons 70-71
  and food digestion control 66, 67
  regulating body temperature 50, 51
  as response to stimulus 74, 78
neuromuscular junctions 72, 102
neurones 41, 70, 72
  in hand withdrawal reflex 78
  vagus 79
neurotransmitters 72, 73, 78
niches 20, 21, 22
nitrate ions 22, 34, 35, 37
nitrifying bacteria 34, 35
nitrogen cycle 3, 34-35
  and deforestation 37
nitrogenous waste products 35, 40, 52-53
nomenclature, biological 18
non-conscious responses (reflexes) 41, 66, 78-79, 104
noradrenaline 51, 78, 79
nucleotides 96
null hypothesis for chi-squared test 9
nutrient cycles 3, 34-35

O
oak trees 29
oestrogen 88, 90, 91, 95, 104
oocytes 86, 91, 95
oogenesis (egg formation) 86, 87
orders 18
ornithine cycle 53
osmosis
  in the kidney 55, 56-57
  in plants 43, 44, 45
ovaries 87, 95
oviducts (fallopian tubes) 87
ovulation, and ageing 104
oxidation reactions 31
oxidative phosphorylation 3, 33
oxygen
  and aerotaxis 80
  from photosynthesis 26
  in respiration 33
  supply to fetus 88, 89, 90
  transport in blood 40, 62-63
oxyhaemoglobin dissociation 40, 62-63
oxytocin 90, 91

P
Pacinian corpuscles 41, 74
pancreas, secretions 48, 64, 65, 67
paneth cells 68
parasympathetic autonomic nervous system 72, 78-79
pea plants
  dihybrid crosses of 8-9
  monohybrid crosses of 6
penicillin resistance 16
pepsinogen and pepsin 64, 67
peptidases 64
percentage cover of plants 20-21
peristalsis 66
pesticide resistance 16, 17
pH
  of blood 62
  and food digestion 64, 65, 67
phenotypes 4
  adaptation of 10
  frequency changes in 14
  influences on 92, 93
  and Mendelian inheritance 6, 7, 8
  variation of 10, 92
phenotypic ratio 6, 8-9
phloem 42, 44, 45
phosphate ions, limiting plant growth 22
phosphocreatine 103
phospholipids 96
phosphorylase enzymes 49
phosphorylation 3, 30, 32, 33, 103
photolysis, of water 26
photosynthesis 2, 26-27
  and carbon cycle 34
  and deforestation 37
  energy transfer in 26, 28
  and gas exchange 59
  and respiration in plants 58, 59
phototaxes 80
phyla (*singular* phylum) 18
phylogenetic classification 19
pigments in cone cells 76
pituitary gland 91, 94-95, 104
pituitary growth hormone (PTH) 94, 95
placenta 88-89, 90, 91
plants
  adaptation 22, 40, 46, 47, 61
  classification 19
  interspecific competition 22, 24-25
  mesophytes 47, 61
  and nitrogen fixation 34
  percentage cover 20-21
  as primary producers 22, 28-29
  respiration 58, 59
  succession of 24-25
  transpiration 40, 44-47
  water and ion transport in 40, 42-43
  xerophytes 40, 47
  see also leaves; roots
plasma 54, 55, 56, 62, 90
podocytes 54
polar bodies 87
polarisation of cell membranes 70
polygenic variation 12, 92
polypeptides, digestion of 64, 65, 67
populations of species
  isolated 15
  size and diversity 22, 36, 37
  size estimation of 21
  stability 16
  variations 2, 10-11
positive feedback 67, 91
postsynaptic membranes 72-73
potassium ion channels 70, 71, 73
potometers 46
predation, and population growth 22
pregnancy 90-92, 98, 99
pressure receptors 41, 74, 79
presynaptic membranes 72
primary producers
  plants as 22, 28-29
  see also plants
progesterone 88, 90, 91, 95
prokaryotic cells 19
prolactin 90, 91
protease 87
proteins
  biuret test for 97
  carrier proteins 48, 55, 68, 70-71, 89
  for cell growth and division 94
  changes during mutation 13
  digestion and absorption 64
  from codominant alleles 7
  and nitrogen cycle 34, 35
  as nutrients 96, 99

as respiratory substrates 31, 96, 98
shape and structure 49, 55, 66, 74, 89, 101
as specific receptors 48, 49, 66, 74, 89, 101
protoctists 19, 58, 65
proximal tubule 54, 55, 56
puberty 94, 95, 98
Punnett squares 8
pupa (chrysalis) 69
pupil, focusing 79
pyramids, ecological 29
pyruvate 32

R
radioactive tracer 45
random sampling 11, 21
Ratchet Mechanism 101
receptors 41, 74-75
heat 50, 78
light, rod and cone cells 41, 76-77
pressure 41, 74, 79
specific *see* specific receptors
stretch 66
recessive alleles 4, 6, 7, 8, 14
recommended daily amounts 99
red blood cells 62-63
reduced coenzymes
in photosynthesis 2, 26
in respiration 3, 32-33
reduction reactions 31
reflexes 41, 66, 78-79, 104
relay neurones 78
renal (Bowman's) capsule 54
reproduction *see* sexual reproduction
reproductive isolation 15
reptiles 50
respiration 3, 30-31
adaptation 58, 60, 61
aerobic 3, 26, 32-33, 60, 103
anaerobic 31, 32, 103
and carbon cycle 34
fetal needs 90
and gas transport 40, 58-59, 62-63
of plant cells 58, 59
rate 94
roles of nutrients in 96
and temperature 24
and water loss 60, 61
respiratory quotient (RQ) 31, 98
respiratory substrates 3, 31, 52, 98
responses to stimuli
kineses 80
nerve impulses 74, 78
non-conscious responses (reflexes) 41, 66, 78-79, 104
taxes 80
resting potential 70, 71
retina 75, 77
Rhizobium 34, 35
rhodopsin 75
ribulose bisphosphate (RuBP) 27
rickets 97
rod cells 41, 75, 76-77
roots 42, 43, 44, 47
rough endoplasmic reticulum 68
ruminants 65

S
saliva 64, 66
sampling 11, 21
sand dune succession 24
saprophytic nutrition 34
sarcomeres 100
Schwann cells 70
secretin 67
seminiferous tubules 85
senescence *see* ageing
sensitivity of rod and cone cells 76-77
sensory neurones 78
Sertoli cells 85, 86
sex-linked characteristics 7, 93, 95, 99
sexual reproduction 4, 14, 84, 85-87
organs 93, 95
shivering 51
shrubs, succession of 24, 25
sickle-cell anaemia 13, 17
sieve elements in phloem 42
silk moths 80
Simpson's Index 23
sinatrial node 79
skeletal (striped/striated) muscle 84, 100-103
skin
receptors in 74, 78
and temperature regulation 50-51, 78
and water loss 60
Sliding Filament Hypothesis 101, 102
sodium chloride transport in kidney 56-57
sodium ion channels 70, 71, 73, 74
speciation 2, 14-15
species 2, 14
abundance 22
classification 18-19
diversity 22-23
new 15
specific receptors 48, 49, 72, 95
spermatozoa (sperm) 85, 86
formation (spermatogenesis) 85, 86, 95
spinal reflexes 41, 78
spiracles 59, 61
standard deviation and standard error 11
starch 48, 64, 97
stimulus
and action potential 71, 74
directional 80
stomata 43, 46, 47, 58, 61
stretch receptors 66
stroma 27, 87
succession 2, 24-25
sugars 2, 27, 96, 97
see also glucose
summation of depolarisations 73
surface area
and diffusion 58, 61, 68, 88
and transpiration rate 47
survival 14, 16-17, 80
suspensory ligaments in eye 75
sweating 51
sympathetic autonomic nervous system 66, 72, 78-79
symplast pathway of water transport 43, 44
synapses 41, 72-73

T
taxa (*singular* taxon) 18
taxes (*singular* taxis) 41, 80
taxonomy 18-19
temperature
and conduction of nerve impulses 71
and ecosystems 24, 37
fetal 90
and nutrient cycles 34, 37
regulation in the body 40, 50-51, 90
and respiration 24
and sperm development 85
and transpiration rate 46
testes (*singular* testis) and testosterone 85, 95
thermoreceptors (heat receptors) 50, 78
thermoregulation in the body 40, 50-51, 90
Thyroid Stimulating Hormone (TSH) 94
thyroxine 94-95
tracheal system, in insects 59, 61
transamination of amino acids 96
transects 20-21
transpiration 40, 44-45, 46-47
trapping methods 21
trees, 23, 24-25, 29, 36-37
trichromatic theory of colour vision 76
triglycerides 31, 65
triose phosphate 27
triplets of bases in DNA 12-13
trophic levels 28
tropical rain forests 36, 37
tropomyosin and troponin 102
trypsinogen and trypsin 64, 67

U
ultrafiltration in kidney 54
umbilical cord, vein and arteries 88, 89
urea 52, 53, 55, 65, 89
uric acid 52
urine 40, 53, 54-55, 56-57, 90
uterus, in pregnancy and birth 90-91

V
vacuolar pathway of water transport 43, 44
vagus nerve/neurones 66, 79
variation in populations 2, 10-11, 12-13
in behaviour 80
vasa efferentia (*singular* vas efferens) 85
vasa recta 57
vasoconstriction/vasodilation 50, 51, 79
vegetarian diets 99
villi (*singular* villus) 68
viruses 19, 89
vision 75-77
vitamins 89, 97

W
warfarin resistance 16, 17
waste products
and energy transfer 28
from respiration 33
nitrogenous 35, 40, 52-53
transport across placenta 89, 90
water
balance in body 40, 56-57
and cohesion tension hypothesis 44-45
from humus 24
loss and its limitation 40, 46, 52, 60-61
photolysis 26
potential
and osmosis 43, 44, 45, 55, 56-57, 96
and transpiration 46
transport in plants 40, 41-42
waxy cuticles
in insects 59
on leaves 46, 47, 58, 61
weight loss diets 99
woodlice 80
worms 63

X
X and Y chromosomes 7
xerophytes 40, 47
xylem 40, 42, 43, 44-45

Z
zygotes 4, 6, 14, 84, 87, 88